E. EHLERS   T. KRAFFT

Understanding the Earth System
Compartments, Processes and Interactions

Springer

Berlin
Heidelberg
New York
Barcelona
Hong Kong
London
Milan
Paris
Singapore
Tokyo

Eckart Ehlers   Thomas Krafft (Eds.)

Understanding the Earth System

Compartments, Processes and Interactions

In Collaboration with:
C. Moss (linguistic editing)

With 80 Figures and 13 Tables

 Springer

Prof. Dr. Eckart Ehlers
Dr. Thomas Krafft
German National Committee on Global Change Research
Walter-Flex-Strasse 3
53113 Bonn
Germany

We gratefully acknowledge the generous support of the Strukturförderungsgesellschaft Bonn/Rhein-Sieg/Ahrweiler mbH for the printing of this publication. The conference "Understanding the Earth System: Compartments, Processes and Interactions" held in Bonn in November 1999 was part of the Bonn Science Festival (Wissenschaftsfestival Region Bonn).

ISBN 3-540-67515-9 Springer-Verlag Berlin Heidelberg New York

Library of Congress Cataloging-in-Publication Data
Die Deutsche Bibliothek – CIP-Einheitsaufnahme
Understanding the earth system : compartments, processes and interactions / ed. by
E. Ehlers & T. Krafft on behalf of the National Committee on Global Change Research in
collab. with: C. Moss. – Berlin ; Heidelberg ; New York ; Barcelona ; Budapest ; Hong Kong ;
London ; Milan ; Paris ; Singapore ; Tokyo : Springer, 2001
    ISBN 3-540-67515-9

Springer-Verlag Berlin Heidelberg New York
a member of BertelsmannSpringer Science + Business Media GmbH

http://www.springer.de

© Springer-Verlag Berlin Heidelberg 2001
Printed in Germany

Production: PRO EDIT GmbH, 69126 Heidelberg, Germany
Cover design: design & production, Heidelberg, Germany
Cover figure: Martin Cassel-Gintz and Hans-Joachim Schellnhuber
Typesetting: Mitterweger & Partner, 68723 Plankstadt, Germany
Printed on acid-free paper   SPIN: 10745351

Preface

This volume includes revised versions of most of the presentations made at the International Conference "Understanding the Earth System: Compartments, Processes and Interactions" held on November 24–26, 1999 in Bonn. The Conference was organized by the German National Committee on Global Change Research as part of the Bonn Science Festival 1999–2000. The Bonn Science Festival (Wissenschaftsfestival Region Bonn) was organized and funded by sfg Strukturförderungsgesellschaft Bonn/Rhein-Sieg/Ahrweiler mbH. The generous support for organizing the conference and printing this volume by sfg is gratefully acknowledged. Additional financial and organizational support for separate workshop sessions and publications have also been provided by the German Federal Ministry for Science and Research, BMBF and Germany's major research funding agency, Deutsche Forschungsgemeinschaft.

The editors wish to gratefully acknowledge the help, advice and especially patience of many individuals who have contributed to this volume. The contributions are intended to document the debate on crucial issues of the emerging concept of earth system science and to stimulate the necessary scientific discussion. While every effort has been made on the part of the editors to ensure consistency in terminology, style and methods of quotation, the variety of contributors has inevitably resulted in certain discrepancies.

Bonn, February 2001 E. EHLERS
T. KRAFFT

Contents

List of Contributors

SURINDER AGGARWAL
University of Delhi
Department of Geography
Delhi 110007
India

MARTIN CLAUSSEN
Potsdam Institute for Climate Impact Research
P.O. Box 601203
14412 Potsdam
Germany

PETRA DÖLL
University of Kassel
Center for Environmental Systems Research
34109 Kassel
Germany

JAMES C. I. DOOGE
University College Dublin
Center for Water Resources Research
Earlsfort Terrace
Dublin 2
Ireland

THOMAS E. DOWNING
University of Oxford
Environmental Change Institute
1 a Mansfield Road
Oxford OX1 3SZ
Great Britain

ECKART EHLERS
University of Bonn
Institute of Geography
Meckenheimer Allee 166
53115 Bonn
Germany

ANDREAS M. ERNST
University of Freiburg
Department of Psychology
Niemensstraße 10
79085 Freiburg
Germany

MARTIN EXNER
University of Bonn
Institute of Hygiene
Sigmund-Freud-Straße 25
53105 Bonn, Germany

PIERRE FRIEDLINGSTEIN
LSCE
Unité mixte 1572 CEA-CNRS
91191 Gif sur Ivette
France

HARTMUT GRASSL
Max Planck Institute for Meteorology
Bundesstraße 55
20146 Hamburg
Germany

MARTIN GROSJEAN
University of Bern
Institute of Geography
Hallerstraße 12
3012 Bern
Switzerland

DANIEL HILLEL
University of Massachusetts and the Hebrew University
Currently: Center for Environmental Studies
P.O. Box 585
37105 Karkur
Israel

ARJEN HOEKSTRA
International Institute for Infrastructural,
Hydraulic and Environmental Engineering
2601 DA Delft
The Netherlands

THOMAS HOFER
Forest Conservation Research and Education Service (FORC)
FAO Rome,
Via delle Terme di Caracalla
00100 Rome
Italy

CARLO JAEGER
Potsdam Institute for Climate Impact Research
P.O. Box 601203
14412 Potsdam
Germany

FREDRICK K. KARANJA
Department of Meteorology
University of Nairobi
P.O. Box 30197
Nairobi
Kenya

MOHAMED SAÏD KARROUK
Centre de Recherche de Climatologie
University Hassan II
P.O. Box 8220 Oasis
Casablanca
Morocco

J. C. KATYAL
National Academy of Agricultural
Research Management, Hyderabad, India
Currently: University of Bonn
Center for Development Research
Walter-Flex-Straße 3
53113 Bonn
Germany

STEPHAN KEMPE
University of Technology Darmstadt
Institute for Geology and Paleontology
Schnittspahnstraße 9
64287 Darmstadt
Germany

THOMAS KISTEMANN
University of Bonn
Institute of Hygiene
Sigmund-Freud-Straße 25
53105 Bonn
Germany

GERNOT KLEPPER
University of Kiel
Institute of World Economics
Düsternbrookerweg 120
24105 Kiel
Germany

THOMAS KRAFFT
University of Bonn
Institute of Geography
Meckenheimer Allee 166
53115 Bonn
Germany

PETER S. LISS
University of East Angelia
School of Environmental Sciences
Norwich, NR4 7TJ
Great Britain

HUBERT MARKL
President of the Max Planck Society
Hofgartenstraße 8
80539 Munich
Germany

LUIS J. MATA
University of Bonn
Center for Development Research
Walter-Flex-Straße 3
53113 Bonn
Germany

WOLFRAM MAUSER
University of Munich
Institute of Geography
80333 Munich
Germany

Bruno Messerli
University of Bern
Institute of Geography
Hallerstraße 12
3012 Bern
Switzerland

Michel-H. Meybeck
University of Paris
Laboratory for Applied Geology/SISYPHE
Vi Case, Tour 26, 5ème
4, Place Jussieu
75252 Paris
France

Lautaro Núñez
Instituto de Investigaciones Arqueológicas y Museo
Universidad Católica del Norte
Casilla 17
San Pedro de Atacama
Chile

Max Pfeffer
Cornell University
Rural Sociology Department
Warren Hall Ithaca
NY 14853
USA

Christian Pfister
University of Bern
Institute of History
Dept. for Economic
Social and Environmental History
Unitobler
3000 Bern 9
Switzerland

Jorge A. Ramirez
Colorado State University
Civil Engineering Department
Fort Collins
CO 80523 – 1372
USA

Mohamed A. Salih
Politics of Development
Institute of Social Studies
P.O. Box 29776
2502 LT The Hague
Netherlands

Hans-Joachim Schellnhuber
Potsdam Institute for Climate Impact Research
P.O. Box 601203
14412 Potsdam
Germany

Wolfgang Seiler
Fraunhofer Institute of Atmospheric Environmental Research
Kreuzbahnstraße 19
82467 Garmisch-Partenkirchen
Germany

Nick van de Giesen
University of Bonn
Center for Development Research
Walter-Flex-Straße 3
53113 Bonn
Germany

Paul L. G. Vlek
University of Bonn
Center for Development Research
Walter-Flex-Straße 3
53113 Bonn
Germany

Rüdiger Wolfrum
Max Planck Institute
for Comparative Public Law and International Law
Im Neuenheimer Feld 535
69120 Heidelberg
Germany

Fred M. Zaal
AGIDS
Universiteit van Amsterdam
Nieuwe Prinsengracht 130
1018 VZ Amsterdam
The Netherlands

PART I

Panorama:
The Earth System: Analysis from Science
and the Humanities

I

Understanding the Earth System –   1
From Global Change Research to Earth System Science

Eckart Ehlers* · Thomas Krafft*

There is no doubt that "Global Change" and its scientific analysis and interpretation are on the forefront of international research efforts. Since the detection of global warming, first signs of world-wide melting of ice-masses and glaciers, indications of sea-level rises and/or the depletion of the atmospheric ozone-layers, increasing numbers of scientists – meteorologists, physicists, atmospheric chemists, oceanographers and others – have devoted their research to the solution of these and related problems. Global change research and its development over the last 20 or 30 years are testimony not only to the almost unbelievable broadening, widening and deepening of research themes, but also – and likewise – to a shift of scientific paradigms. As a matter of fact: the title of this conference "Understanding the Earth System – Compartments, Processes and Interactions" and the publication of its proceedings are part of this development.

Climate Change and Global Environmental Change (GEC)

The last two decades have seen major changes not only of research foci, but also of changing perceptions and approaches toward their analyses. The organization of global change research, its growing number of international and interdisciplinary research groups, the scientific expansion of originally focussed research programs, their increasing differentiation in scope and objectives: all this reflects the growing complexity of global change research. Not surprisingly, it was the climate research community that led the way. They pointed to those first indications of a dramatic climatic change, demonstrated – among other signals – by the following observations:

- tropical ocean surface temperatures increased by 0.5 °C;
- increase in tropospheric water vapour concentrations above the tropics;
- rises in the amount of bound heat, released in the mid-layers of the tropical troposphere;
- growing temperature gradients between equator and lower latitudes;
- correspondingly increased mean wind velocities.

Global change research therefore started off as climate research and the climate research community contributed the first important scientific breakthroughs in

* e-mail: sekretariat.nkgcf@uni-bonn.de

the understanding of climate variability stimulating further research (cf. contri-butions by Grassl and Klepper in this volume). Most of the scientific results were achieved in the context of the World Climate Research Program (WCRP), the old-est of the international global environmental change research programmes. WCRP's research agenda was designed around four major aims: "(a) to improve our knowledge of global and regional climates , their temporal variation, and our understanding of the responsible mechanisms; (b) to assess the evidence for sig-nificant trends in global and regional climates; (c) to develop and improve physi-cal mathematical models capable of simulating and assessing the predictability of the climate system over a range of space and time scales; (d) to investigate the sensitivity of climate to possible natural and man-made stimuli and to estimate the changes in climate likely to result from specific disturbing influences". The different IPCC reports published in the last decade are decisively based on research findings from WCRP and document the progress of the programme.

From observation and interpretation of climate change to attempts of compre-hensive impact assessments was the obvious next step. Geologists, hydrologists, biologists as well as representatives of other terrestrial and marine geo-sciences joined the chorus of those who tried to understand the identified climate change indicators. Their focus, however, was less on the physical climate system, but on the biogeochemical impacts and processes in both marine and terrestrial envi-ronments. With other words: the International Geosphere-Biosphere Program (IGBP), created in 1985 by the International Council of Scientific Unions (ICSU), added new dimensions to global change research by focussing on climate change impacts on terrestrial and marine ecosystems, at the same time, however, asking increasingly for the earth surface's role and human impacts in their importance for environmental change. IGBP's original mission statement and goal reads as follows: "To describe and understand the interactive physical, chemical and bio-logical processes that regulate the total Earth system, the unique environment that it provides for life, the changes that are occurring in this system, and the manner in which they are influenced by human actions".

IGBP research and its focus on the biogeochemical processes that respond to and – at the same time – cause climate changes therefore marks the beginning of global *environmental* change research, environment understood as the physi-cal environs and ecosystems of our globe. With such an ambitious and enlarged research agenda much more than before integration and interaction between different and so far predominantly disciplinary orientated research communities became indispensable. Testimony to this growing interdisciplinarity is the well-known Bretherton diagram (see fig. 1). It indicates necessities, potentials and limits of joint and cross-disciplinary research. At the same time, however, it gives clear hints as to the desirable, however, yet to be defined clear participa-tion of the humanities and social sciences in regard to global environmental research.

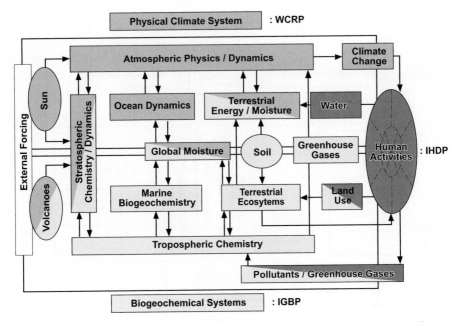

Fig. 1. A modified "Bretherton diagram" highlighting some of the linkages between social systems, biogeochemical systems and the physical climate system. [Source: IGBP.]

Social Sciences and Global Environmental Change Research

Speculations, theories and scientific facts about societal causes of global environmental changes are as old as science itself and surely not less fundamental in their findings. With regard to the theme and topic of our conference and this publication, however, it is decisive and unquestioned that modern global environmental change research has been and still is dominated by the natural sciences. This holds true in spite of the fact that natural science has, from the very beginning and again and again, demanded social science's engaged participation in global environmental change research. It holds true, however, also in spite of the fact that the social sciences have developed a research agenda of their own. Nevertheless, they have failed for a long time to join forces with natural sciences.

As early as 1989, the International Social Science Council (ISSC) created the so-called Human Dimensions of Global Environmental Change Program (HDP) in order to complement WCRP and IGBP. In a preliminary statement, aims and goals were described as follows: "Human activities that interact with the Earth's natural systems are driven by three fundamental factors: the number of human beings and their distribution around the globe: their needs and desires – as conditioned by psychological, cultural, economic, and historical factors – which provide their motivations to act; and the cultural, social, economic, and political structures and institutions and norms and laws that shape and mediate their

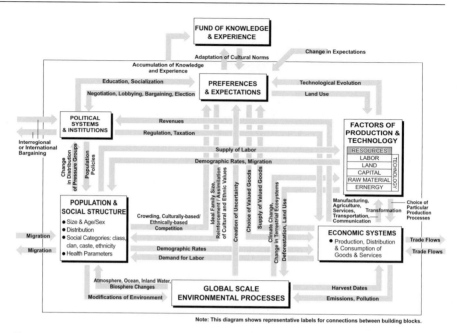

Fig. 2. Human Dimensions of Environmental Change. Social Progress Diagram. [Source: Consortium for International Earth Science Information Network (1992).]

behaviour. Consequently, research on the human dimensions of global environmental change must consider human behaviour in relation to three broad themes: the social dimensions of resource use; the perception and assessment of global environmental conditions and change; and the impacts of local, national, and international social, economic, and political structures and institutions on the global environment. Such research must be conducted at all geographical scales and should include the past as well as the present and the future" (Jacobson – Price 1991, p 19).

The comparison of the Bretherton diagram with that of HDP (see fig. 2) reveals the basic differences of scientific self-understanding, research approaches and designs as well as theoretical background and methodological implications. Though this is not the place to discuss these differences in detail, it is obvious that aims and scope of natural versus social science research in the field of global environmental change research seem basically different and neither scientifically comparable nor methodologically coherent.

It took some years until social science research and the HDP underwent a re-evaluation and re-structuring. Under joint sponsorship of both ISSC and ICSU since 1995, the new International Human Dimensions of Global Environmental Change Program (IHDP) was launched in 1996, with the following mission statement: "IHDP is an international, interdisciplinary social science programme to promote and coordinate research aimed at describing, analysing and understanding the human dimensions of global environmental change, i.e.: the way people and societies contribute to global environmental change; the way global environmental change effects people and societies; and ways and means for people and societies to mitigate and adapt to global environmental change."

The focussed redefinition of IHDP coincided with a growing awareness of the international science community of potentially greater impacts of people and societies on our natural environments than hitherto anticipated. Growing insights into human alterations of the earth's ecosystems and their devastating impacts on atmosphere, biosphere, geosphere and hydrosphere added new dimensions to the explanations of global environmental changes. The detection of a "human domination of the earth's ecosystems" (Vitousek et al. 1997) caused concerns: could it be that the global society and its energy demands are a more powerful driving force of global environmental changes than nature itself?

In recent years this question of "nature dominated" versus "human dominated" global environmental change has gained momentum (see also Messerli in this volume). And this question, increasingly raised by natural scientists themselves, is the more important since it refers not only to the basic fundamentals of global environmental changes and their causes, but also to the methodological questions of scales: are *global* observation systems sufficient for analyses of change and predictions for the future or do we need regional, or even: local observations? Very probably, W.E. Easterling's affirmative statement is generally acceptable (Easterling 1997). Its acceptance, however, implies acknowledgement of the fact that human activities are an essential component in regard to any assessment of global change processes. The question remains of how to incorporate natural and social science and how to achieve science-orientated theoretical and methodological commonalities!

From Global Environmental Change Research to Earth System Analysis

As early as 1986, H. Markl has pointed to the fact that we are living in an "anthropozoic" geological era in which humans – individuals as well as societies – tend to exert a much bigger influence on terrestrial changes and on marine life than nature sensu stricto (see also the contribution of Markl in this volume). Quite recently, Crutzen and Stoermer (2000) have suggested to use the term "anthropocene" for the current geological epoch emphasizing the still growing impact of human activities on earth and atmosphere. Though their suggestion to set a starting date for the "anthropocene" in the later part of the 18th century might be debatable, there is no doubt that mankind has emerged as a significant force in the earth system affecting or altering key processes of the global environment. The identification of the human fingerprint on the state of the global atmosphere, the world oceans, and the land surface has brought about the challenge for science and society to define and effectively implement policies for sustainable development that are based on a profound understanding of the dynamic links and interdependencies between natural systems and human actions.

The attempts to include "man", i.e. societal impacts as part of a comprehensive and system-orientated analysis of the overall earth-system marks the most recent

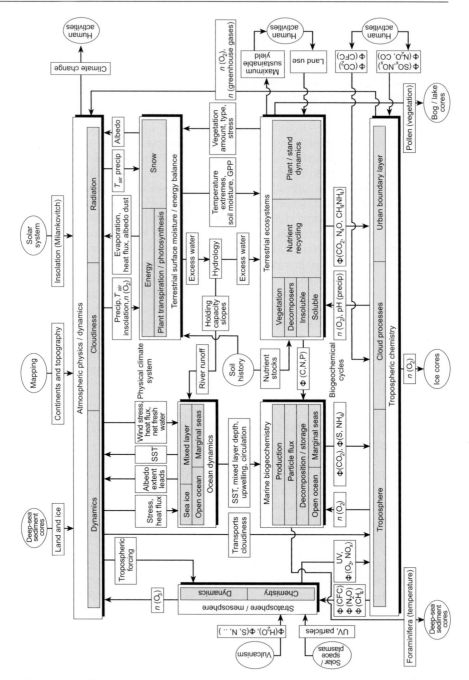

Fig. 3. A simplistic conceptual model of the planetary machinery. [Source: Schellnhuber (1999).]

development; as elaborated in H.-J. Schellnhuber-V. Wenzel's programmatic publication on "Earth System Analysis" (1998) which constitutes at the same time a dramatic plea for cross-disciplinary integration of research and for the development of a "science of sustainability".

Earth System Analysis as conceptualized by Schellnhuber (cf. contribution in this volume) intends to combine the traditional disciplinary-sectoral or "reductionist" approaches with new simulation or modeling approaches to provide a better understanding of the earth system. However, there is also an increasing number of other recent and inspiring publications pointing in the same direction. Theoretically, at least, the gap between natural sciences on the one hand and the social sciences/humanities on the other hand seems to narrow. Practically, however, methodological deficiencies and seemingly invincible inconsistencies remain. It is therefore not surprising that also the most recent attempt to develop a kind of earth system-approach (simplistic, as it is called and as it may be!) still fails to convincingly incorporate the human causes and effects in regard to global environmental changes in an integrative-holistic way (see fig. 3).

Earth System Science: Research Problems and Necessities

This admittedly crude review of the development from global change research to earth system science and the introduction into the following collection of articles on "Understanding the Earth System: Compartments, Processes and Interactions" would be incomplete if one would not point explicitly to basic research problems and address the need for urgent solutions. Blue-eyed as such a statement may be: the integration of natural and societal factors into an overall and comprehensive earth system approach seems precondition for any successful attempt in direction of a comprehensive earth system science. Being well aware of the controversial discussions about "holistic-integrative" research designs and problems of their implementation, there seems to be a general consensus about the fact that earth system science without the human factor is an incomplete systemic approach. Recent discussions (for a useful collection of articles on theoretical and methodological problems of integrative environmental research one may point to the recent edition of Daschkeit-Schröder 1998) seem to indicate that, at present, solutions in regard to coherent research strategies of both natural and social sciences are not in sight. There are at least three dilemmas that need to be solved in any future pursuit of a truly integrative and holistic attempt to bridge theoretical and methodological gaps between disciplines in the common attempt to understand the earth system and its functioning mechanisms:

Dilemma 1: Objects of research
Dilemma 2: Compatible methodologies
Dilemma 3: Problems of scale

Objects of research: It is almost trivial to state that natural sciences are mainly experimental disciplines which – based on mathematics – deal with the material realities of our earth system and which try to understand and explain their laws.

Social sciences, on the other hand, are those that deal with human society, societal groups or individuals in their relationship to others or to societal institutions. Material and/or cultural goods and values are expressions of their common contents and theoretical foundations. How can such basic differences and almost juxtaposed approaches be combined under one umbrella?

Referring to the afore-mentioned diagrams (see fig. 1–3) it is rather difficult to imagine any reasonable form of "cohabitation" of those projections. Is it possible at all? A somewhat simplistic answer could be: Surely not on the basis of traditional disciplinary self-understanding and/or jealously defended scientific turfs. On the contrary: interdisciplinary research on the global environment and its changes cannot proceed from disciplinary problem-definitions or scientifically subjective self-defined research agendas. They have to be derived from objective problems and policy issues. Or – as Michael Redclift has called it: "knowledge cannot be divorced from the uses to which it is put." (Redclift 1998).

In a recent publication the Board on Sustainable Development (BSD) of the US – American National Research Council has suggested – among others – to "develop a research framework that integrates global and local perspectives to shape a 'place based' understanding of the interactions between environment and society" and " to initiate focused research programs on a small set of under-studied questions that are central to a deeper understanding of interactions between society and the environment" (BSD 1999, pp 10–11).

Two examples of practical, user-orientated and problem-solving research approaches may demonstrate ways and methods in which cooperation between natural and social sciences is almost preconditions for any reasonable research strategy. The first example refers to the much discussed "concept of syndromes", i.e. functional patterns (syndromes) that are unfavourable and characteristic constellations of natural and civilisational trends and their respective interactions, overcultivation of marginal lands, overexploitation of ecosystems, environmental damages due to large-scale development projects – to name just a few examples! – demand place-based integrative research approaches (cf. also Schellnhuber in this volume).

Another approach may be the development of jointly designed and formulated research projects where a specific problem is at the very base of such an attempt. The AQUA/GLOWA project, developed by the German National Committee on Global Change Research (see fig. 4) may be an appropriate example. Problems of availability, quality and allocation (AQUA) of water in different ecological and socio-economic settings are the focus of a set of research projects, jointly designed by social and natural scientists and to be implemented on the basis of comparable catchment areas along an ecologically differentiated gradient from Europe to tropical Africa.

Compatible methodologies: The problem of bridging the gap between the different methodologies is, of course, and again closely connected with the object as well as with the aim and scope of the research agenda. Bruno Latour, author of the well-known study "We have never been modern", argues that "we have developed three distinct approaches to talking about our world: naturalisation, socialization and reconstruction" (Latour 1993). Especially reconstruction, a widening gap

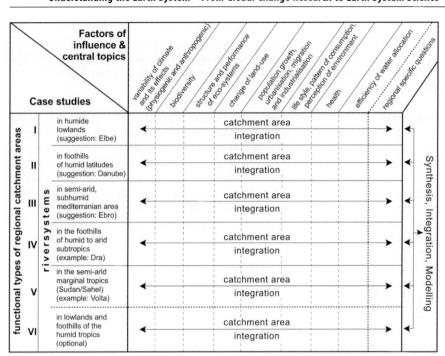

Fig. 4. Integration Matrix. Water: Availability, Quality and Allocation. [Source: National Committee on Global Change Research (2000).]

between the "nature pole" and the "society pole" since the age of enlightenment and the massive development of "quasi objects" have greater contributed to the asymmetries of science: "The more the quasi-objects multiply, the greater grows the distinction between the poles" (Latour 1993, p 58).

One of the main deficiencies in the development of GEC-research over the last 12 to 15 years has been – as indicated – the hesitance of social scientists to participate more actively in the promising and future-orientated endeavours of this new and huge research field. Whatever reasons for this far-reaching abstention there might be, it is obvious that large segments of the social science community appear to be neither interested nor knowledgeable about GEC-research. Obvious is also, however, that on human dimensions focused GEC-research has to leave trodden disciplinary paths and to develop new methodologies that enable the social sciences to feed the results of their research into local, regional, and/or global models and scenarios and thus to respond to the expectations of the natural sciences. Their questions are manifold. Experience shows that attempts of "socialising the pixel" as well as "pixelizing the social" are beginning to bear rich fruit (cf. articles in Livermann et al. 1998). These attempts show that social science can contribute to modelling. On the other hand: integration of human dimensions research is not only a one-way-street. It also needs the critical self-evaluation of the natural sciences in regard to the validity of their global models and predictions. The fact that meteorologists, oceanographers, climate physicists and others question these validities and call for regional scale models with the inclusion of human dimension parameters is promising in regard to the future GEC-research.

Problems of scale (spatial and temporal): "Global Change" – this very term speaks for itself and – at the same time – indicates the direction out of which this specific kind of research has developed. It have been the physical science perspectives that have dominated the early phases of GC-research and its policy-contexts. In line with natural scientists' predominant self-understanding according to which science is timeless and spaceless and knowledge a process of gradual accretion on a linear uni-dimensional basis, it is not surprising that global change research started off with a global perspective and global models. Social sciences, in contrast, are neither timeless nor placeless. Therefore, it may be understandable that early and positivistic driven GC-research of the natural sciences is characterised by global top-down models, while the more interpretative social science approaches towards GEC-research do not only proceed from local or regional analyses, but also tend to follow a bottom-up approach.

It has taken some time to recognize the importance and validity of a combination of both top down global models versus bottom up local/regional or sectoral models of GEC-research. The research design of the joint IGBP-IHDP program on Land-Use and Land-Cover-Change (LUCC) is one of those promising examples of up- and down-scaling of research results (see HDP-IGBP 1995, esp. figs. 6 and 7). The fact that especially the IGBP-community will focus its future research on medium-scale regional models is another indication that the problems of scale will be harmonized in the future. And that global approaches are important for human dimensions research and climate change policies is obvious. On the other hand: historical, religious, cultural or socio-economic diversity of humankind contradicts uniform explanations and/or policy-orientated recommendations.

On the whole it is probably unrealistic to expect sudden or basic shifts of theoretical-methodological paradigms in the fields of global environmental change research and earth systems analysis. Differences in objects, theories and methodologies in its implementation have grown apart since the enlightenment and the specialisation of scientific disciplines. On the other hand there is growing insight into the complexity of the earth system and the intricate interdependencies of natural and societal causes of its changes Therefore, new holistic and integrative approaches towards their analysis and understanding may not only be necessary but indispensable.

Understanding the Earth System: A Plea for Integrating Research, Knowledge and Action

Returning to the outset of this introductory article, it should be repeated that global change studies, global environmental change research and earth systems analysis are on the forefront of international research efforts – and it is very likely that this will remain for the years to come. As indicated before there is, however, also a growing insight into the holistic complexity of this type and topic of research – and into the necessity to combine research with action. The plea to develop a "science of sustainability" and to create a science-based pursuit towards sustainable management of the earth system coincides with the necessities of integrated, maybe even holistic approaches to both research and action. The present publication should be seen as a contribution to such attempts.

The ongoing debate on identifying pathways toward sustainable development and the recurrent call for "global environmental management" or "earth system management" underlines the compelling need to reduce uncertainties in our understanding of changes in the global environment. "Efforts to manage humanity's interactions with the global environment [...] will almost certainly be partial, contentious, and prone to failure. But the management of global environmental risk is also what an increasing number of political leaders, advocacy groups, scientific experts, and international organizations find themselves doing, not uncommonly with the best of motivations and the greatest of trepidation" (Clark et al. 2000, p 2). Scientific knowledge has therefore to contribute reliable guidance for society's decision on appropriate pathways toward sustainable development by providing evidence and answers which help to identify certainties and reduce uncertainties in our understanding of global environmental issues. The further development and future success of international agreements and cooperation require that society not only understands, but also acquiesces to the consequences of environmental protection. This again depends on a growing understanding of the Earth System not only among scientists, but in particular by society (cf. Wolfrum in this volume).

It is this strategic role that also creates the need to reconsider the relationship between science and society along the lines of Lubchenco's demand for a new social contract for science while we are entering the century of environment. Such a contract would have to be "predicated on the assumption, that scientists will (i) address the most urgent needs of society, in proportion to their importance; (ii) communicate their knowledge and understanding widely in order to inform decisions of individuals and institutions; and (iii) exercise good judgement, wisdom, and humility" (1998, p 495). Gibbons' call for a new social contract between science and society to ensure reliable but even more "socially robust knowledge" goes further in addressing the importance of "producing" science both in a transparent and participative way. "Under the prevailing contract, science was left to make discoveries and then make them available to society. A new contract will be based upon the joint production of knowledge by society and science. A new social contract will therefore involve a dynamic process in which the authority of science will need to be legitimated again and again" (1999, p C84).

During the last two decades GEC-research has significantly contributed to identify environmental threats driven by a variety of human activities and to advance our understanding of the earth's biological, physical and chemical systems. The research efforts that lead to the comprehensive framework for controlling worldwide emissions of ozone destroying substances as documented in the Montreal Protocol are but one example of these important scientific contributions. Others include the success to provide reliable information on El Niño phenomena, the documentation of evidence of past changes of Earth's environment based on palaeoclimatic research or the progress in large-scale modelling of terrestrial and marine ecosystems (cf. contributions by Liss and Claussen in this volume). Building on the basis of these achievements research efforts need now to concentrate on more integrated and place-based approaches to analyse already identified or new emerging threats to the global environment and to assess available options for dealing with them. Focusing on the critical scientific questions that need to be resolved to ensure a better understanding and to develop successful environmental policies will transform GEC-research toward a more comprehensive earth system science. However, in the decades to come disciplinary basic research will continue to be the backbone of our research also providing a basis of knowledge for the likelihood of unexpected developments.

While disciplinary basic research provides the fundamental knowledge and the necessary technological capabilities for systemic and integrative analysis, the "grand challenges" of global environmental change can only be addressed through the co-operation and interaction of both natural and social sciences. The division of GEC-research during its formative years into four major components corresponds to the institutional arrangements made by ICSU, ISSC and WMO through the establishment of the four international programmes (WCRP, IGBP, IHDP, and DIVERSITAS). Each of these programmes was intended to be complementary to the others. In recent years joint WCRP/IGBP and IGBP/IHDP projects have been successfully launched or are in the process of implementation. However, further integration will be necessary to provide comprehensive answers to the central issues of global environmental change. Future integrative programmes have also to give increased attention to both human dimensions of global environmental change and to biodiversity research. Support of this scope must be based on the development of coherent, focused and integrative as well as innovative new research projects in which natural and social science research communities work together on problem-solving and politically relevant issues.

The decision of the Scientific Committee of IGBP to base the future development of the programme in the first decade of the 21st century on the principals of *integration, interdisciplinarity* and *a systems approach* reflects the new orientation toward earth system science. This ambitious approach is based on the earlier experience of GEC-research that "understanding components of the Earth System is critically important, but is insufficient on its own to understand the functioning of the Earth System as a whole" (Moore 2000, p 1). The new research approach leading from "linear, pollution-pipe" models to integrated system approaches, from focused sub-system science to integrated earth system science (Steffen 1999) will enhance the need for a closer collaboration among the four international GEC Programmes to build a common scientific foundation. The focus on three cross-cutting issues of major societal relevance is anticipated to help develop a closer collaboration across disciplines, programmes and regions. The *Global Carbon Cycle, Water Resources* and *Food and Fibre* constitute the themes of these cross-cutting initiatives with other major issues such as *Health and Environment* under consideration.

The International Conference "Understanding the Earth System: Compartments, Processes and Interactions" organized by the German National Committee on Global Change Research was designed to provide a forum for the necessary dialogue between researchers both from science and the humanities on global environmental change research. The conference aimed to promote a scientific discourse on the pros and cons of integrative-holistic versus disciplinary research and to contribute to the debate on a comprehensive "synthesis" of GEC-research. The following presentation of the major scientific contributions to the conference in this volume follows the original structure of the conference. Three major themes were covered: (1) Panorama: The Earth System – Analysis from Science and the Humanities, (2) Focus: Water in the Earth System – Availability, Quality and Allocation in Cross-disciplinary Perspectives, and (3) Perspective: Advancing our Understanding – Reductionist and/or Integrationist Approaches to Earth System Analysis. In context with theme 2 "Water in the Earth System" – as mentioned before one of the cross-cutting issues of GEC-research – a couple of workshops and breakout groups were organized focussing on achievements and challenges of water research in the context of Water Quality and Health Risks, Urban Thirst and Water Conflicts, Modelling Water Availability: Scaling Issues, Precipitation Variability and Food Security, Water Deficiency and Desertification, and Interdisciplinary Perspectives on Fresh Water: Availability, Quality, and Allocation. The major findings and discussions of these workshops and breakout groups are summarized and documented in the appendix to this volume. The different contributions combined in this volume give vivid evidence that GEC-research is a policy driven, problem-orientated, and interdisciplinary field of applied science in which the co-operation between natural and social sciences are precondition for scientific success and achievements. In addition, as we move on from GEC-research toward earth system science we will participate in restructuring science and contribute to a new "consilience" i.e. unity of knowledge.

References

Becker E, Jahn Th, Stiess J, Wehling P (1997) Sustainability: A Cross-Disciplinary Concept for Social Transformation. Paris (UNESCO: Management of Social Transformations/ MOST Policy Papers 6)

Board on Sustainable Development (1999) Our Common Journey. A Transition towards Sustainability. National Academic Press, Washington, D.C.

Clark WC, Jäger J, van Eijndhoven J (2000) Managing Global Environmental Change: An Introduction to the Volume. In: Clark WC, Jäger J, van Eijndhoven J, Dickson NM (eds) Learning to Manage Global Environment Risks: A Comparative History of Social Responses to Climate Change, Ozone Depletion and Acid Rain. MA: MIT Press, Cambridge

Consortium for International Earth Science Information Network (eds) (1992) Pathways of Understanding. The Interactions of Humanity and Global Environmental Change. University Center, The Consortium for International Earth Science Information Network, MI, USA

Crutzen PJ, Stoermer EF (2000) The "Anthropocene". IGBP Newsletter 41: 17–18

Daschkeit A, Schröder W (eds) (1998) Umweltforschung quergedacht. Perspektiven integrativer Umweltforschung und -lehre. Springer, Berlin – Heidelberg – New York

Easterling WE (1997) Why regional studies are needed in the development of full-scale integrated assessment modelling of global change processes. Global Environmental Change 7: 337–356

Gibbons M (1999) Science's new social contract with society. Nature Supplement to 402: C81–C84

HDP-IGBP (1995) LUCC – Land Use and Land Cover Change. Science/Research Plan Stockholm-Geneva. IGBP Report No. 35 / IHDP Report No. 7

IGBP-IHDP (1999) LUCC-Land Use and Land Cover Change. Implementation Strategy. Stockholm-Bonn. IGBP Report 48 / HDP Report 10

Jacobson H, Price MF (1991) A Framework for Research on the Human Dimensions of Global Change. IHDP Report Series No 1

Latour B (1993) We have never been modern. Harvard Univ Press, Cambridge/Mass

Livermann D, Moran EF, Rindfuss RR, Stern PC (eds) (1998) People and Pixles – Linking Remote Sensing and Social Science. National Academic Press, Washington D.C.

Lubchenco J (1998) Entering the Century of the Environment: A New Social Contract for Science. Science 279: 494–479

Markl H (1986) Natur als Kulturaufgabe – Über die Beziehung des Menschen zur lebendigen Natur. DVA, Stuttgart

Messerli B et al. (2000) From nature-dominated to human-dominated environmental changes Quaternary Sciences Review 19: 459–479

Moore B (2000) Sustaining Earth's life support systems – the challenge for the next decade and beyond. IGBP Newsletter 41: 1–2

NRC (1994) Science Priorities for the Human Dimensions of Global Change. NRC Committee on the Human Dimension of Global Change / Commission on Behavioural and Social Sciences and Education. National Academic Press, Washington, D.C.

Redclift M (1998) Dances with wolves? Interdisciplinary research on the global environment. Global Environmental Change 8: 177–182

Schellnhuber HJ (1999) 'Earth System' analysis and the second Copernican revolution. Nature Supplement to 402: C19–C23

Schellnhuber HJ, Block A, Casel-Gintz M, Kropp J et al. (1997) Syndromes of Global Change. GAIA 6: 19–34

Schellnhuber HJ , Wenzel V (eds) (1998) Earth System Analysis. Integrating Science for Sustainability. Springer Verlag, Berlin

Steffen W (1999) Global Change Science in the next century , a personal perspective. IGBP Newsletter 40: 4–5

Vitousek PM, Mooney HA, Lubchenco J, Melillo JM (1997) Human Domination of Earth Ecosystems. Science 277: 494–497

Wilson EO (1998) Consilience. The Unity of Knowledge. New York

CHAPTER 2

Earth System Analysis and Management

H. J. SCHELLNHUBER*

Understanding the Earth System is neither a necessary nor a sufficient condition for its management. On the one hand, the present civilizatory interference with the atmosphere demonstrates that *bad* management, at least, driven by the opportunistic interests of zillions of individual actors unaware of the long-term consequences, is feasible at a planetary scale. On the other hand, we can change our ways for the better even without perfect (predictive) knowledge about the functioning of the global machinery including the human factor. We need to know, however, *something*, and the production and use of this something is the topic of my contribution.

Discovering Planet Earth

There are two things in the universe which are utterly special (complex, fragile, and beautiful), namely our colourful planet as floating in dark cold space, and the human body in the incipient stage of maturity. Figure 1 contrasts these two astonishing achievements of natural self-organization in pictorial form.

The human body has been used as a powerful metaphor for the analysis of the planetary make-up by Jim Lovelock and his scientific companions in their pioneering effort to establish a "geophysiologic theory" (for reviews, see, Lovelock 1991, Schneider and Boston 1993, Lenton 1998). In fact, thorough studies of our compositional and metabolic physique may teach Earth System scientists many useful lessons (Volk 1998). But even from a purely epistemological point of view, it is worth while taking a look back on the bizarre story of the unravelling of the mysteries of the human body.

Western medicine and its heuristic strategies have been dominated – if not tyrannized – by the Hippocratic school for more than 2000 years. Diagnostics and therapy of the deplorable patients were almost exclusively based upon "humoral pathology", interpreting diseases as manifestations of accidental deviations from a hypothesized natural harmony of body liquids like blood and bile. Galenos, the Hippocatic plaster saint, was, for instance, convinced that wounds can be healed by pus stimulation (removal of *materia peccans*). In the course of

* e-mail: john@pik-potsdam.de

Fig. 1. a Planet Earth as seen from outer space. **b** Adam and Eve as imagined by Albrecht Dürer.

time, this dogma may have caused more human casualties than all the wars instigated by megalomaniacs.

A paradigm shift was initiated (yet not completed) in the middle of the last millennium by two scholars that number among the outstanding pioneers of the Enlightenment: Vesalius and Harvey. Andreas Vesalius, a Belgian physician of German extraction, published his main oeuvre, *De Humanis Corporis Fabrica Libri Septem*, in 1543 – the same year that *De Revolutionibus Orbium Coelestium* by Nicholas Copernicus appeared. As Copernicus drew the anatomy of the heavenly bodies, Vesalius visualized for the first time the true structure of the human body and created the modern science of anatomy. The seven volumes were not just a cognitive but also an artistic world sensation, as the analysis was lavishly illustrated by Tizian using Campagnola landscapes as backgrounds (see Fig. 2). The *Fabrica* is often called "the greatest medical book ever written".

The second giant shattering Hippocrates's empire was the British physiologist William Harvey, who unravelled the precise mechanisms driving the blood circulation (*Exercitatio Anatomica De Motu Cordis Et Sanguinis In Animalibus*, published in 1628), and even laid the foundations for heart surgery in the 20th century.

Modern medical science has come a long way since the days of Galenos and company, and it is still a long way from home: defeating cancer, deciphering the immune system, exploring the encephalon, and, ultimately, understanding what the soul is all about – these challenges will keep researchers busy in the decades to come. And yet we can be confident that almost all the secrets of the human body will be revealed in the not-too-distant future.

Today, around the turn of the 2nd millennium, we are witnessing the early stage of another discovery story, comparable in importance and character to the

Fig. 2. One of Tizian's "musclemen" visualizing Vesalius's anatomical wisdom.

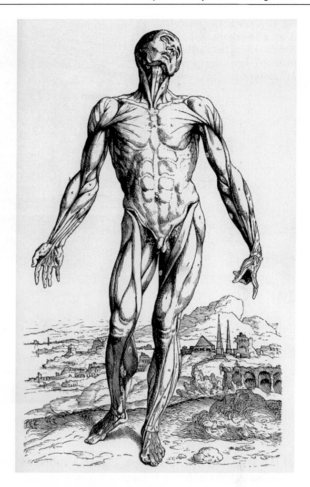

one pioneered by Vesalius and Harvey. This time the object of scientific scrutinization is "Gaia' s body", i.e. the physique of the ecosphere that supports, inter alia., the species *homo sapiens sapiens*.

In a recent essay for the Millennium Supplement of *Nature* (Schellnhuber 1999), I have interpreted this research venture as the "Second Copernican Revolution": While the first Copernican revolution put the Earth in its correct astrophysical context and opened humanity's view into the surrounding cosmos, the second one will allow us to look back on our planet, perceiving a unique complex dynamic entity that has been driven far away from physicochemical equilibrium by self-generated biological forces in an autopoietic process. We do perceive, in particular, a system that may be nature's sole successful attempt at creating in our Galaxy a life-form that is appropriately tailored for explaining the possibility of such an entity – and of itself. The implicit option for self-control will be a focus

Fig. 3. A tale of two revolutions. **a** The shock of the Enlightenment; **b** "Earth-System" diagnostics in the 21st century.

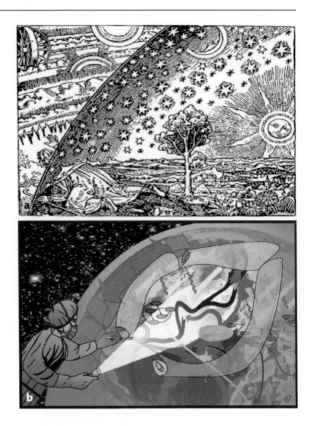

of the following considerations. Figure 3 contrasts the two dialectically related paradigm shifts in cartoon form.

Macroscoping Planet Earth

The Enlightenment, which firmly established the scientific method for treating reality over a time span of eventually four centuries, was based, literally, on the processing of light: The Great Copernican Revolution used telescopes to blow up minute fire flies on the nocturnal heavens to stately meteorites and planets; the cannons that finally shot to pieces the dogmas of the Hippocratic school were microscopes, which even tracked down the bacterial agents causing those diseases that plagued the ancient world more than famine or tempests. Both success stories were due to the same technical principle, namely the magnification of light rays through ingenious combinations of curved lenses.

For the inspection of the planetary machinery, which embraces all kinds of global-scale processes from plate tectonics to ionosphere pulsation, it is necessary to compress information rather than to amplify it. This means that instru-

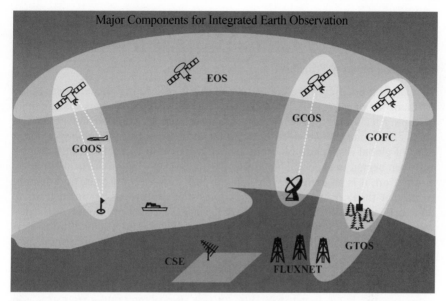

Fig. 4. Crucial elements of envisaged Earth Reconnaissance System integrating remote sensing and ground truth across all relevant sub-spheres of our planet.
Abbreviations: CSE – Continental Scale Experiments; EOS – Earth Observing System; FLUX-NET – Global Network of Flux-Measurement Stations; GCOS – Global Climate Observing System; GOFC – Global Observation of Forest Cover; GOOS – Global Ocean Observing System; GTOS – Global Terrestrial Observing System.

ments are needed that reduce the virtual size of the Earth System (or of significant parts of it) to a degree that allows top-down contemplation ("Ganzheitsbetrachtung"). The French scientist and philosopher Joël de Rosnay has introduced the term "macroscope" to denote such instruments: "Il nous faut donc un nouvel outil ... cet outil, je l'appelle le *macroscope* ... Le macroscope filtre les détails, amplifie ce qui relie, fait ressortir ce qui rapproche. Il ne sert pas à voir plus gros ou plus loin. Mais à observer ce qui est *à la fois* trop grand, trop lent et trop complexe pur nos yeux" (de Rosnay 1973).

There are three major ways to build a macroscope for watching our planet and the footprints of humanity on it.

The Bird's-Eye Principle

The first option is to achieve "holistic" visions by monitoring the Earth from a distance (satellites, space shuttles, planes and other remote-sensing devices) or by vast arrays of observation stations or sensors that cover crucial components of the planetary physique (Sellers et al. 1995, Kibby 1996, GTOS 1998, Running 1998, King and Greenstone 1999, Justice et al. 1999, Defries and Belward 2000). The international scientific community plans to combine several of those macroscopes to establish an almost complete Earth reconnaissance in the near future. Figure 4 sketches the basic structure of the overall system.

The Digital-Mimicry Principle

Simulation modelling represents a canonical way of creating virtual copies of the Earth, or major parts of it, in cyberspace. We will elaborate on this specific macroscoping technique in sect. 4, but anticipate here its main advantage and its main disadvantage: The electronic chimaeras representing spatiotemporal sections of global reality can be exposed to all sorts of imaginable perturbations – and even beaten to experimental death – without jeopardizing the true specimen in question. It is an extremely challenging task, however, to secure and demonstrate that those chimaeras are *valid* copies that capture all pertinent traits of their originals. Even the most advanced general circulation models for atmospheric dynamics (Stouffer et al. 1999, Voss and Mikolajewicz 2000, Gordon et al. 2000, Pope et al. 2000) do not stand all the tests of reproducing the observed (past and present) climatic behaviour.

In view of such problems the Japanese initiative for constructing and running an "Earth Simulator" appears to be a bit premature: This hyper-computer is being constructed near Yokohama; it will cost almost 500 Million US $ of investments, and 10 megawatt electric power under standard operational conditions. 5120 high-speed processors will combine to accomplish a peak computing performance of 40 teraflops, while the central storage unit will guarantee a 10 terabyte memory.

One of the ambitious goals to be met on this unprecedented technological basis is to implement and evaluate, by 2005, a coupled atmosphere-ocean general circulation model (AOGCM) of 1 km resolution! Taroh Matsuno, the project director, is confident that this novel macroscope will bring about a quantum leap in predictive capacity for all sorts of geophysical phenomena, ranging from ENSO events to volcano eruptions.

Yet simulating Planet Earth needs more than number crunching and software engineering: due to multiple non-linear interactions within the real system, the virtual representations of certain processes have to be extracted with utmost precision from theoretical principles and observational data sets whereas other mechanisms may require only crude and qualitative digital images. There is no way to avoid scientific trial-and-error schemes here, but note that the empirical testing of many hypotheses in this field may be either unfeasible or irresponsible (Schellnhuber 1998a). This dilemma leads us directly to the third major option for macroscope building.

The Lilliput principle

An attractive strategy for studying the operational niche of our planet seems to be "nano-technology", which rebuilds the original or relevant parts thereof at a much smaller scale. This approach could solve, in principle, some of the main problems associated with the first two macroscoping approaches: On the one hand, the specimen in question reflects by construction the real physical features of the parent system. On the other hand, the Lilliput copy is sufficiently insignificant to allow for sacrifice on the altar of scientific curiosity. Cosmology and elementary-particle physics actually employ this heuristic principle excessively. Gigantic particle colliders generate, for instance, "little bangs" that reproduce conditions equivalent to those prevailing in the first moments of the creation of the universe. A recent achievement was the production of an extremely confined quark-gluon plasma, representing the most exotic state of matter ever realized deliberately by humanity.

Compared to these accomplishments, nano-reproduction of the planetary machinery is still in an embryonic stage. The most spectacular attempt made so far was the "Biosphere II" venture, enacted in the Sonora Desert some forty miles north of Tucson, Arizona. On September 26, 1991, four male and four female "bionauts" were "sealed inside a 3.15 – acre glass and steel enclosure in which they would have to recycle their water, grow their own food, and generate their atmosphere through plants. The purpose of this 2-year mission was to answer the question of whether a complex, materially closed, energetically and informationally open ecosystem including several biomes could persist and support human life for that period" (Bechtel et al. 1997). The overall setting of this Earth macroscope is depicted in figure 5.

A somewhat detailed discussion of the usefulness and (im)possibility of Earth-System reproduction in miniature is given in the forthcoming report of the German Advisory Council on Global Change (2000). What should be mentioned here is the fact that Biosphere II became a splendid failure. The tiny ecocosmos behaved completely differently from what had been expected and, therefore, raised a plethora of salient questions for complex-systems analysts. The present state of the art does not allow to answer the question why Biosphere II went astray the way it did.

Most probably, the future of Earth watching through macroscopes will be dominated by employing intelligent and effective combinations of the three approaches described, particularly of the first two. Planetary monitoring by remote sensing and a worldwide net of *in situ* measurement arrays will be perpetually complemented and synchronized by data models. Today, already, we are able to track down hurricanes over their complete life cycle in slow motion and every detail, if we like. The electronic reconnaissance reveals, i.a., how a number of these singular meteorological phenomena build up via convergence and convection off the coast of West Africa, how they gain momentum on their journey westwards across the Atlantic, how they get reflected and attenuated by the American landmass, and how they travel back towards Europe as mundane depressions. This is reminiscent of watching the motion of bilestones across the human body using modern "mediscopy".

The perspectives for Earth macroscopy are nicely illustrated by the Brazilian mega-project SIVAM: this is supposed to become the "biggest ecological monitoring system in the world", a technological complex consisting of hundreds of remote-sensing devices and thousands of computers, costing altogether approximately 1.4 billion US $. The objective of SIVAM is the complete on-line observation and analysis of the entire Amazon region. As a side-effect, the system is expected to hunt up illegal gold washers and loggers. "Big brother" is watching himself...

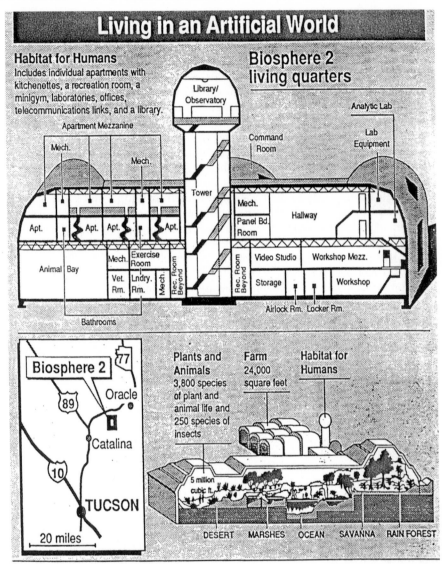

Fig. 5. The layout of Biosphere II.

Defining the Earth System

In the previous sections, we have used the notion "Earth System" as a buzz word without specifying the meaning of this increasingly popular term. Let us try to catch up on a no-nonsense definition now.

At the highest level of abstraction, the make-up of the Earth System $\mathcal{E}$ can be represented by the following "equation":

$$\mathcal{E} = (\mathcal{N}, \mathcal{H})$$

$$(a, b, c, \ldots) = \mathcal{N}\,\mathcal{H} = (A, S) \tag{1}$$

$$S = (B, V, O).$$

A detailed interpretation of this formal structure is given in Schellnhuber (1998a); here we restrict ourselves to sketching the crucial features only. First of all, Eq. 1 expresses the elementary insight that the overall system in question consists of two main components, namely the *ecosphere* $\mathcal{N}$ and the *human factor* $\mathcal{H}$. $\mathcal{N}$ consists, in turn, of an alphabet of intricately linked planetary sub-spheres a (atmosphere), b (biosphere), c (cryosphere), and so on. This is the entity described with the metaphor "Gaia" in geophysiologic theory (Lovelock 1991). The human factor is even more subtle: $\mathcal{H}$ embraces the "physical" sub-component A (*anthroposphere* as the sum of all individual human lives, actions and products) and the "meta-physical" sub-component S reflecting the emergence of a "*Global Subject*". This subject is a self-organized cooperative phenomenon, a self-conscious force driving global change either to sustainable trajectories or to self-extinction. For the time being, let us just mention that S may be decomposed into three constituents, namely B (the global "brain"), V (the global value system, or "soul"), and O (the executive "organs" of the Global Subject). We will return to this "metaphysics" at the end of the section.

All the relevant aspects defining the character of the Earth System are, in principle, also reflected by Drake's equation (see e.g. Drake and Sobel 1992, Dick 1998). This is an attempt to quantify the contemporary number of technical civilizations in our galaxy, N_{CIV}. The number is evidently (still) larger than zero. Drake's equation reads as follows:

$$N_{CIV} = N_{MW} \times f_P \times n_{CHZ} \times f_L \times f_{CIV} \times \delta. \tag{2}$$

Here N_{MW} denotes the present number of hydrogen-burning main-sequence stars in the Milky Way, f_P the fraction of those stars that have planets, n_{CHZ} the average number of planets that lie within the "continuously habitable zone", f_L the fraction of those habitable planets that develop a full ecosphere, f_{CIV} the fraction of those biospheres that bring about a technical civilization, and, finally, δ the average ratio of technical-civilization lifetime to hydrogen-burning main-sequence stars age. All these definitions lead us deeply into the thriving field of "astrobiology". Obviously, the estimation of the factors determining N_{CIV} in eq. 2 becomes more and more difficult – if not hopeless – as we move from left to right: a tentative quantification of δ involves actually a complete theory of the sustainability of advanced planetary civilizations under all sorts of extrinsic and intrinsic destruction forces!

It seems reasonable, however, to try to evaluate the product of the first four factors on the right-hand side of eq. 2. This product, i.e.,

$$N_{GAIA} = N_{MW} \times f_P \times n_{CHZ} \times f_L, \tag{3}$$

represents the approximate number of ecospheres ("Gaias") that have come into being on suitable planets orbiting suitable stars in our galaxy. The key factor in eq. 3 is n_{CHZ}, which refers to the abundance of planets that reside for a sufficiently long time (10^9 years, at least) in the so-called habitable zone (see e.g. Hart 1979). This zone is defined, to a first approximation, by the band of planet-star distances that warrant moderate surface temperatures – and, as a consequence, the existence of liquid surface water – on the orbiting body. The overall concept has been considerably refined by Kasting and a number of co-workers (see, for instance, Kasting 1997).

Recent studies by a research team at the Potsdam Institute (including myself) have pushed this astrobiological analysis even further (Franck et al. 1999, Franck et al. 2000). The investigation is based upon a geobiodynamic model for long-term Earth-System evolution that accounts for all pertinent processes involved in the generation of photosynthesis-driven life conditions. A crucial precondition for habitability within the framework of this analysis is the existence of *plate tectonics* on any planet appropriately positioned in its solar system. One of the immediate applications of that model is a filtering scheme that allows to identify those individuals among the crowd of recently discovered extrasolar planets (Schneider 1999), which are candidates for supporting a fully-fledged ecosphere. It can be demonstrated, for instance, that the famous "MACHO–35" planet (see the references in Franck et al. 1999) is definitely not suitable for extraterrestrial life. In fact, habitability would require that the red "planetary line" in figure 6 crosses the green domain.

Another important application of our modelling approach is the estimation of N_{GAIA}, in which we are most interested in the context of Earth System analysis. Combining rigorous simulation results and subjective expert judgement in the sense of Bayesian reasoning, we find that

$$N_{GAIA} = 2.4 \times 10^6. \tag{4}$$

Fig. 6. Shape of the habitable zone (green shading) in the mass-time plane for an Earth-like planet at the probable MACHO–35 distance R = 2 AM from the central star. For the details we refer to Franck et al. (1999). If MACHO–35 is assumed to have about one Earth mass, the calculated central-star mass of 0.3 solar units is much too small to warrant habitability.

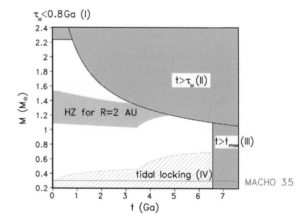

This represents a quite big and viable ecosphere population in our galaxy through time. The number of contemporary technical civilizations in the Milky Way, N_{CIV}, may be much smaller, however. The present estimates range from 10^6 to 10^{-15} (Walter 1999). This would mean that the "Search for Extraterrestrial Intelligence" (SETI) is probably a futile enterprise. For sister Gaias, out there beyond the limits of our solar system, are generally not accompanied by a cultural factor based on the higher forms of self-organization.

This is obviously different in the case of the Earth System. The ecosphere, $\mathcal{N}$, and the physical component of the human factor, $\mathcal{A}$, are coevolving at an ever increasing pace and form an intricately linked dynamical complex since the onset of "Global Change" (that is, since the middle of the 20th century). Representing the state of the ecosphere and the anthroposphere by the vector variables N and A, respectively, we can – symbolically – express this co-dynamics by the following systems equation:

$$\frac{dN}{dt} = F_1 (N, A; t),$$

$$\frac{dA}{dt} = G_1 (N, A).$$

(5)

Here F_1 is some generalized Lagrangian function which explicitly depends on time (due to extrinsic forcing through astrophysical or geophysical perturbations like cosmic dust clouds or reversal of Earth's magnetic field), while G_1 is the counterpart governing the development of the anthroposphere.

A cartoon representation of eq. 5 is given in figure 7.

Note that the present-day quasi-equivalence of the ecosphere and the anthroposphere as physical co-factors of planetary evolution is beyond any doubt: Humanity now manipulates, e.g., approximately 40 % of the net primary terrestrial production caused by green plants (Wright 1990). Many other evidences for the magnitude of the human impact can be found in the pertinent lit-

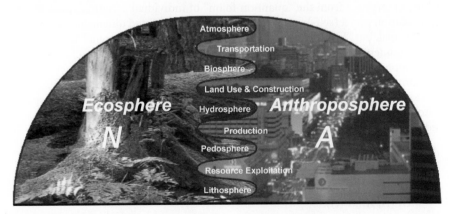

Fig. 7. The inter-penetration of $\mathcal{N}$ and $\mathcal{A}$ in the early "anthropocene".

Fig. 8. The material and the immaterial components of the Earth System, which reside in orthogonal, yet mutually constitutive, dimensions.

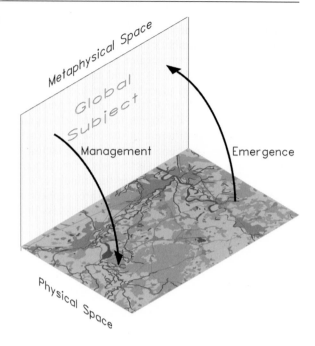

erature (see e.g. Schellnhuber and Kropp 1998). The most conspicuous example of all is the "ozone-hole story" which demonstrates that subtle non-linear synergisms triggered by minute anthropogenic admixtures to the natural atmosphere can bring about catastrophic deviations from the equilibrium mode of ecosphere operation. There is no need to repeat all the bleak details about contemporary global environmental change here.

Let us instead move on to formally close the fundamental analysis of the Earth System: this closure is achieved by the Global Subject, S, introduced above, whose emergence from the "quantum foam" of individual activities turns $\mathcal{N} - \mathcal{A}$ co-evolution into a (self-)control exercise of unprecedented dimensions and quality. S is real, but "lives" on a metaphysical space that is "orthogonal" to the physical components of $\mathcal{E}$. This overall make-up of the Earth System is sketched, again in cartoon form, in figure 8.

Before elaborating a bit on S, let us emphasize that the formal representation of Earth-System dynamics is completely transformed by the management aspects associated with the existence of a Global Subject. Instead of eq. 5, we have to consider the control problem:

$$\frac{dN}{dt} = F_2\,(N, A; t; M\,(t)),$$

$$\frac{dA}{dt} = G_2\,(N, A; M\,(t)).$$

(6)

The crucial new factor here is the "external" management strategy M(t), which is selected from a vast pool, $\mathscr{M}$, at the disposal of $\mathcal{S}$ for the sake of steering co-evolution along desired paths. If certain goals are met, the whole exercise may be dubbed "Sustainable Development" (see Sect. 5 and Schellnhuber 1998a). The elements of $\mathscr{M}$ may also be conceived as "geo-cybernetic options" (Schellnhuber and Kropp 1998), embracing, e.g., a well-designed system of global environmental conventions. The evolution functions F_2, G_2 are the counterparts of F_1, G_1 in eq. 5.

The Global Subject, who decides upon the actual strategy for "managing planet Earth" (Clark 1989) is the result of the "fifth step" in the more than four billion years of natural self-organization on *Terra*: "Bacteria, complex cells, multi-cellular organisms, and societies may be the four major transitions in evolution that we know about. But I believe we are now embarking on a fifth transition. *Homo sapiens* is slowly evolving into something akin to a superorganism, a highly-structured global society in which the lives of everyone on the planet will become so interdependent that they may grow and develop with a common purpose. It may take centuries to fully evolve this global creature, but there is no question that it will – only what it will ultimately be like" (Jolly1999). Two keys to the emergence of $\mathcal{S}$ from the physical basis are world-wide transportation and telecommunication which generate, inter alia, thousands of virtual societal super-cells such as personal networks, pressure groups, non-governmental organizations, epistemic communities, and global corporations. Another crucial fac-

Fig. 9. Global Leviathian in the age of *homo trans-sapiens*.

tor is the building and application of macroscopes which allow S to watch itself in action; this self-referentialism will bring about self-consciousness at the planetary scale.

The emergence of S is the birth of a modern "Leviathan", transcending the seventeenth-century imaginations of the English philosopher Thomas Hobbes by many light-years.

Figure 9 provides an allegorical image of the Global Subject in the form of a collage of some constitutive elements.

As a matter of fact, even the "lost continent" – Africa – will be joining the fifth transition towards higher complexity rather soon. "Africa One", the biggest commercial venture in the history of that continent, will create a glass-fibre ring along the entire coast-line, some 39,000 km long. This project will firmly integrate the African countries into the digital revolution transforming all cultures on Earth.

Understanding the Earth's Physique

Let us return, for a while, to more solid scientific ground, namely the material part of $\mathcal{E}$. The analysis of the $\mathcal{N}\text{-}\mathcal{A}$ complex is, in fact, an extremely challenging task in its own right and the basis for any type of planetary management by any sort of global willpower.

There are now four research projects of global design and ambition that strive to meet that challenge: the World Climate Research Programme (WCRP), the International Geosphere-Biosphere Programme (IGBP), the International Human Dimensions Programme (IHDP) and DIVERSITAS, a programme in *statu nascendi* which aims to understand the structure and functioning of biological diversity on our planet. While these programmes will probably carry on in the next five years, a strong movement within the scientific community also asks for organizing, in the not-too-distant future, an integrated venture on "Earth System Science" (ESS) that embraces all crucial aspects of nature-civilization interactions at the global scale as transpiring from regional processes. While even the already available wisdom accumulated by the individual programmes so far may be amplified by a huge factor through intelligent integration, the prospects for genuine transdisciplinary co-operation are even finer:

For some of the natural partners in such an ESS team have come a quite long way and have accomplished amazing things already. This is illustrated in a conspicuous way by the development of IGBP over the last one-and-a-half decades. The overall design of this research programme, especially the systemic allocation of the "Core Projects" and "Task Forces", is sketched in Box 1 (IGBP 1999). This mission statement reveals that IGBP is the first genuinely global and transdisciplinary enterprise in the history of science – as far as its ambitions are concerned, at least.

In fact, many of the goals defined a long time ago in a state of cognitive innocence seem to be achieved or in reach as the programme is heading for its synthesis phase. IGBP has discovered, described and analysed a number of megafluctuations, teleconnections, feedback loops and self-stabilization mechanisms

BOX 1

Objectives and Structure of IGBP. IGBP's goal is to describe and understand the interactive physical, chemical and biological processes that regulate the total Earth System, the unique environment it provides for life, the changes that are occurring in this system, the manner in which they are influenced by human action.

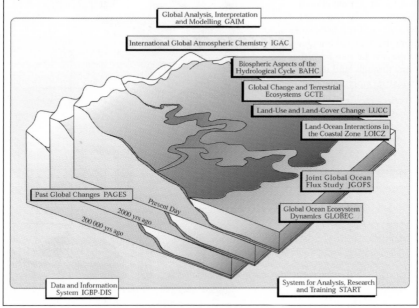

in the entrails of the planetary ecosystem. Several of the subsequent contributions of this book will elaborate on that. For the sake of illustration, let us mention here – in a totally eclectic way – a few highlights from recent core-project studies:

Within the framework of IGBP-PAGES it has become possible to reconstruct, in great detail, global and regional climatic history as far back as 500,000 years ago (Petit et al. 1999). IGBP-JGOFS is more concerned with the present operational mode of the Earth System. It tries, in particular, to solve the puzzle of CO_2 flux across the entire marine environment. Intermediate results indicate that CO_2 is upwelling and leaving the ocean in the subarctic western Pacific in winter, in the Persian Gulf in summer, and west of South America all year round. In contrast, oceanic regions, where warm currents like the Gulf Stream and the Kuroshio are being cooled, take up large amounts of CO_2 (Takahashi et al. 1997). This marine analysis is complemented by IGBP-GTCE activities that scrutinize the CO_2 metabolism of the terrestrial biosphere. Recent findings provide hints that the continental ecosystems may turn into a huge carbon *source* as a result of unabated climatic change some time after 2050. A subtle combination of factors, including the transformation of ecological "disturbance regimes" (fires, storms,

Fig. 10. The terrestrial world as made up from biomes. This geo-physiologic tableau reflects model results (BIOME 1.1, Prentice et al. 1993) for potential ecosystems structure under present climatic conditions. The real distribution significantly deviates from the depicted one due to massive human perturbations through land use, industrial emissions, etc.

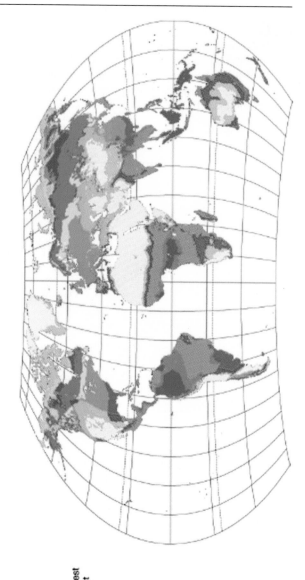

tropical rain forest
tropical seasonal forest
tropical dry forest
xerophytic woods/scrub
hot desert
warm grass/shrub
broadleaved evergreen forest
temperate deciduous forest
cool mixed forest
cold mixed forest
cool conifer forest
cool grass/shrub
S cold deciduous forest
N cold deciduous forest
S taiga
N taiga
S tundra
N tundra
semidesert
ice/polar desert

BIOME 1.1
normal climate

pests, etc.) is responsible for this discomforting prospect (Walker et al. 1999). No less disturbing is a result from IGBP-LOICZ, which focuses on the Earth's coastal zones. Model simulations indicate that crucial mechanisms for coral-reef accumulation in the tropics could be considerably weakened by the anthropogenic modification of the atmosphere's chemical composition (Kleypas et al. 1999).

All these explorative spikes illuminate how a general picture about the physiology and metabolism of the ecosphere is transpiring from IGBP research. This picture still resembles a puzzle with many important pieces missing; however, the fundamental structure determining the systems properties becomes visible through the thinning veils of ignorance. This means, in particular, that the Earth-scientific community is developing a clear conception of the "vital organs" of the planetary machinery – in analogy to the medical discovery story outlined in sect. 1 (Volk 1998). There are many reasons to identify those organs with the archetypes of large-scale terrestrial and marine ecosystems that ultimately emerged from geobiospheric coevolution. Thus "Gaia's body" is governed by the land-based *biomes* (Prentice et al. 1992) and the oceanic *biogeochemical provinces* (Platt et al. 1995). Figures 10 and 11 provide visual impressions about this conceptual decomposition of the ecosphere.

Fig. 11. A section of the marine world as made up from biogeochemical provinces (by courtesy of Ducklow 1999). This partitioning of the oceans reflects especially observations (by, e.g. remote imagery) of ecophysiological parameters. The provinces are determined by geophysical oceanographic and climate forcing; their boundaries vary in space and time.

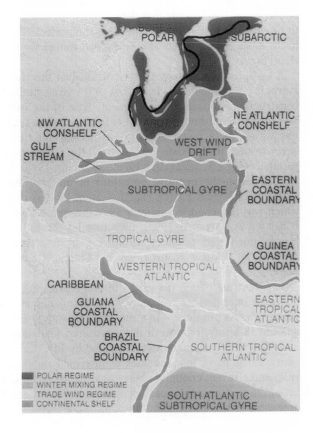

IGBP contributes in many other ways to the unravelling of the mysteries of the ecosphere's anatomy, and thereby pioneers the Second Copernican Revolution. This would be impossible, as mentioned above, without the massive use of macroscopes of all kinds. Very recently, IGBP-GAIM has started the manufacturing of a particularly powerful and effective tool in that methodological ensemble – *intermediate complexity modelling* of the material Earth System. Digital mimicry is generally a most appropriate way of integrating – in a "theatre world" – the main processes and forces that drive and link the $\mathcal{N}$–$\mathcal{A}$ complex. When devising specific models, however, the scientific designers have to resist two fatal attractions:

The first one is over-simplification which tends to produce models that cannot reproduce most of the crucial elements of planetary dynamics like the episodic warming of the tropical East Pacific called "El Niño" (Timmermann et al. 1999). These *tutorial* models (Schellnhuber and Kropp 1998) usually do not even try to simulate the overall topology, let alone the metric proportions, of the original. By focussing on two or three fundamental mechanisms like insolation-vegetation-albedo feedback (see e.g. the "Daisy World" family as reviewed in Lenton 1998), however, they may considerably advance our understanding of non-linear mechanisms that ultimately shape the planetary mode of operation. The second dangerous pull is towards over-sophistication, often resulting in bulky and horrendously expensive models that defy a transparent analysis no less than the reality to be simulated. Let us not forget that the perfect "Earth Simulator" is the Earth itself, more reliable than any *analogical* model we may patch up the Frankensteinian way.

A compromise strategy tries to retain just the right degree of complexity by averaging over the details irrelevant to the issue studied. This can be achieved by constructing reduced-form representations of the full set of dynamic equations according to well-known principles from complex-systems analysis and scientific computing. As far as the atmosphere module (the simplest part of $\mathcal{E}$, by the way) is concerned, there exists, for instance, the option to separate "slow" and "fast" variables (see e.g. Saltzmann 1978, Saltzmann 1988) or to apply certain filtering techniques to the primitve equations of motion (Petoukhov et al. 2000). The GAIM initiative on EMICs (Earth System Models of Intermediate Complexity), which was launched in Summer 1999 through a kick-off workshop in Potsdam (Claussen and Cramer 1999), tries to pilot the modelling community into this largely uncharted territory, promising fresh cognitive pastures. A thorough account on EMICs is provided by Martin Claussen's contribution to this book.

For the time being, it suffices to emphasize one of the incontestable *raisons d'être* for intermediate-complexity modelling, namely the capability of caricaturing the Earth's physique – in a recognizable way – with computing resources that amount to a tiny fraction of those required for comprehensive simulators. This capability is convincingly demonstrated by PIK's home-made EMIC, CLIMBER, which exists in various versions of intermediate complexity. CLIMBER is both valid and efficient enough to serve as a "time machine" that produces virtual vistas of the environmental past, present, and future. The large theatre of operation accessible to this EMIC is outlined in figure 12.

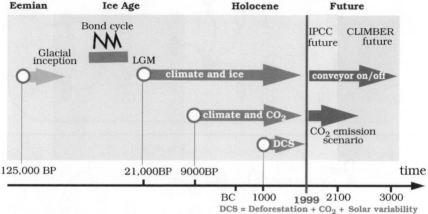

Fig. 12. Schematic view of one of the topical-temporal domains that is currently explored by a specific EMIC, CLIMBER–2, at the Potsdam Institute.

So EMICs may eventually become protagonists in yet another success story of global-change science. We must not conceal the fact, however, that models up and running now are almost entirely restricted to mimicking the dynamics of the ecosphere $\mathcal{N}$. No attempts have been made so far to employ such tools for simulating also the relevant behaviour of $\mathcal{A}$, let alone $\mathcal{S}$. Integrating "at least" the anthroposphere into intermediate-complexity Earth-System modelling would require, first of all, an intermediate-complexity scientific analysis of societal dynamics that harmonizes with the natural-systems treatment. Since the *principia mathematica* for socioeconomic behaviour of individuals, groups and states have not been discovered yet – and probably do not exist in a concise Newtonian form – it is extremely difficult to come up with an appropriate comprehensive theory which would then be reduced to fast simulation algorithms that capture the crucial features of the system in question.

So the right recipe to the modelling of $\mathcal{A}$ may be inverse to the one employed in geophysical mimicry: simulate in a crude *rule-based* way individual actors or compounds of actors in social dynamics, let huge masses of those creatures interact in virtual reality, and run these numerical exercises in large ensembles under stochastic initial and boundary conditions on parallel super-computers. There is some hope that crisp quantitative relationships ("laws of socioeconomic dynamics") will emerge in the "thermodynamic limit", i.e. when the number of rules, actors, and ensembles approach infinity. These laws should express, in particular, the long-term non-equilibrium behaviour of complex anthroposphere entities like national or international markets. That behaviour is not satisfactorily captured by neo-classical macroeconomic models, for instance, although fresh initiatives for "evolutionary" approaches promise to provide some

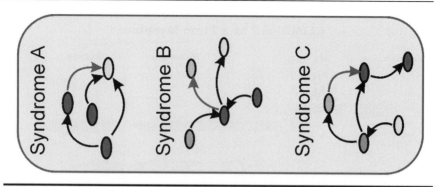

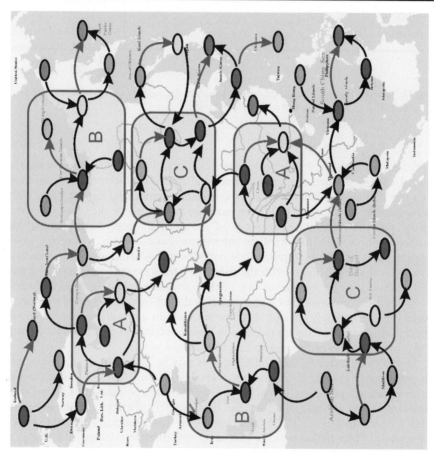

Fig. 13. The "global-change tapestry" as woven from regional cause-effect patterns.

remedy. Many aspects of the topical scientific discourse on socioeconomic dynamics are discussed in the forthcoming book *The Science of Disaster* by Bunde and Schellnhuber (2001).

As long as Earth System science still lacks complementary anthroposphere models that deserve the label "intermediate complexity" something less sophisticated – that still makes sense – has to stand in. Driven by the need to produce policy-relevant results just in time for the successive stages of the international negotiations on the Framework Convention on Climate Change (FCCC), a number of research groups developing "integrated assessment" schemes for optimal climate management (see e.g. the review by Schneider 1997) have realized and partially solved that problem. The socioeconomic modules employed there range from simple Lagrangian optimization algorithms (Nordhaus 1992) over rule-based respectively, intuitive heuristic approaches (see, for instance, Rotmans and de Vries 1997, Alcamo et al. 1998) to brute-force general market equilibrium simulation machines (see e.g. Walker 1997, The Energy Journal 1999) that may be seen as precursors of the desired micro-dynamical multi-actor models sketched in the preceding paragraph.

A novel Earth-System model at the bottom layer of the intermediate-complexity domain is the ICLIPS simulator (Bruckner et al. 1999, Petschel-Held et al. 1999a), which is primarily used as an instrument for integrated assessment of climate change. This simulator tries to incorporate all pertinent elements of ecosphere and anthroposphere dynamics at a comparable reduced-form level, and heavily exploits systematic "knock-out criteria" (tolerability thresholds) via inverse calculation techniques. In other words, this approach does not strive to determine optimal paths for $\mathcal{N}$–$\mathcal{A}$ co-evolution but safe environment and development corridors that avoid potential catastrophe domains. The inversion trick solves at one blow many analytical and computational problems posed by the intricacy of the ecosphere-anthroposphere complex.

There is yet another initiative at the Potsdam Institute that attempts to overcome the cognitive difficulties of Earth System analysis by reconciling the natural and human dimensions in a most unconventional way: this is the so-called "Syndromes Approach" which decomposes the mega-phenomenon "Global Change" into archetypical patterns (Petschel-Held et al. 1999b, Schellnhuber et al. 1997). The syndromes project is based upon the master hypothesis that the web of functional relationships governing the planetary $\mathcal{N}$–$\mathcal{A}$ dynamics is made up of a *finite* set of spatiotemporal recurrent sub-webs of distinct causal topology. This heuristic strategy is conceptualized in Figure 13.

The syndromes concept was pioneered by the German Advisory Council on Global Change (1995, 1997). The overall approach consists of two major steps, namely (i) the identification of the basic syndromes themselves, and (ii) their representation in a way that transcends pure description and even allows for predictive respectively interventional analysis. For the sake of validity, the first – constitutive – step is performed in a two-fold way, i.e. both in a top-down ("complexification") and a bottom-up ("simplification") fashion.

For complexification one starts from the dominant trends of Global Change as unfolding interactively on the fundamental compartment structure of the Earth System (see fig. 14).

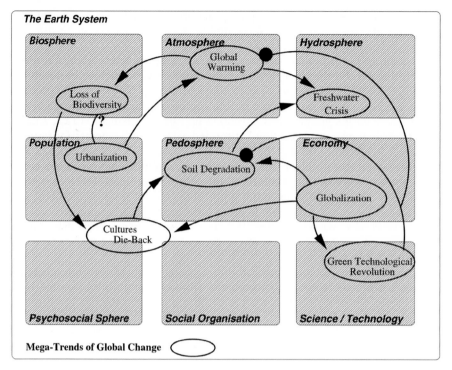

Fig. 14. Mega-trends of Global Change co-evolving across the sub-spheres of $\mathcal{E}$.

On closer scrutinization – resolving the crucial causal relationships in a sufficiently detailed and geographically explicit manner – a consolidated collection of typical degradation patterns emerges (potentially sustainable sub-dynamics are not taken into account for the time being but may play an important role in future research). This cognitive process is depicted in symbolic form in figure 15.

The resulting syndromes taxonomy is summarized in box 2 (see also National Research Council 1999).

BOX 2

Overview of Global Change Syndromes

"UTILIZATION" SYNDROMES

1. Overcultivation of marginal land: *Sahel Syndrome*
2. Overexploitation of natural ecosystems: *Overexploitation Syndrome*
3. Environmental degradation through abandonment of traditional agricultural practices: *Rural Exodus Syndrome*
4. Non-sustainable agro-industrial use of soils and bodies of water: *Dust Bowl Syndrome*

5. Environmental degradation through depletion of non-renewable resources: *Katanga Syndrome*
6. Development and destruction of nature for recreational ends: *Mass Tourism Syndrome*
7. Environmental destruction through war and military action: *Scorched Earth Syndrome*

"DEVELOPMENT" SYNDROMES

8. Environmental damage of natural landscapes as a result of large-scale projects: *Aral Sea Syndrome*
9. Environmental degradation through the introduction of inappropriate farming methods: *Green Revolution Syndrome*
10. Disregard for environmental standards in the course of rapid economic growth: *Asian Tigers Syndrome*
11. Environmental degradation through uncontrolled urban growth: *Favela Syndrome*
12. Destruction of landscapes through planned expansion of urban infra-structures: *Urban Sprawl Syndrome*

"SINK" SYNDROMES

13. Singular anthropogenic environmental disasters with long-term impacts: *Major Accident Syndrome*
14. Environmental degradation through large-scale diffusion of long-lived substances: *Smokestack Syndrome*
15. Environmental degradation through controlled and uncontrolled disposal of waste: *Waste Dumping Syndrome*
16. Local contamination of environmental assets at industrial locations: *Contaminated Land Syndrome*

For simplification, i.e. the complementary bottom-up derivation of the syndromes, one scales up appropriate relational observations from the fund of local and regional case studies that has been amassed by the natural and the social sciences (and, in particular, by geography) over the last decades (see fig. 16).

Box 3 adds some flavour to the conceptual outlines of the syndromes approach drawn so far. As an illustration, the so-called Sahel Syndrome is introduced and analyzed in some detail there.

As for the second major step mentioned above, i.e. formal modelling of the syndromes identified, a rigorous and precise mathematical representation is currently out of reach. Smart *semi-quantitative* methodologies like those employed to an increasing extent in artificial-intelligence research, however, are promising candidates for doing the job. In particular, qualitative differential equations (see, for instance, Kuipers 1994) are powerful tools for integrating all sorts of knowledge – whether exact, soft, subjective or fragmentary. The resulting models cannot reproduce or predict correct numerical values for the object variables to be simulated, but they can explore the qualitatively distinct classes of solutions of the underlying dynamics, and can track down the evolution of these solutions in

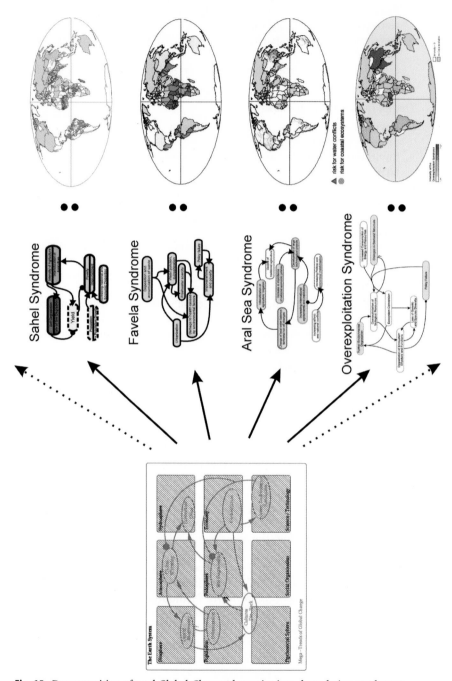

Fig. 15. Decomposition of total Global-Change dynamics into degradation syndromes.

BOX 3

Core dynamics, global intensity map and global disposition (susceptibility) map for the Sahel Syndrome (consult Cassel-Gintz et al. 1997 for a full interpretation)

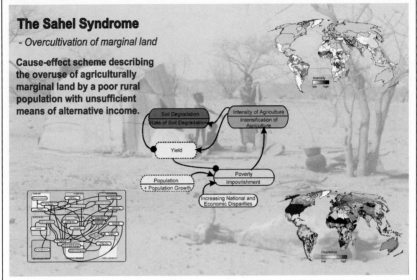

topological time. This semi-quantitative representation of real systems is, of course, underdetermined, so instead of a single trajectory, a solution tree, bifurcating into the future, is generated. In figure 17 we demonstrate this behaviour for a tutorial qualitative differential equations simulation of the Sahel Syndrome.

Apart from their application to syndromes analysis, semi-quantitative techniques do not yet play a big role in Global-Change science. These techniques,

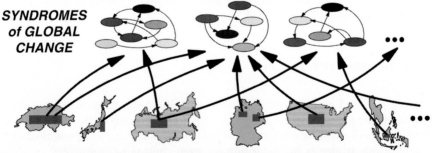

Fig. 16. Distillation of degradation syndromes from geographically explicit information.

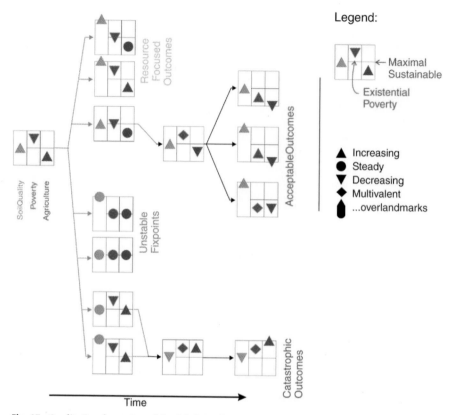

Fig. 17. Qualitative dynamics of the Sahel Syndrome as resulting from a simplified cause-effect web (for details, see Schellnhuber et al. 2000).

however, may hold the key to large domains of the intermediate-complexity terri-tory, and can become invaluable for Earth System analysis. We can be confident that the next two decades will bring about a dramatic boom of research and accomplishments in this very field.

Managing the Earth System

Having reviewed now our state of knowledge about the Earth System and the prospects for cognitive quantum leaps – what are we going to do with this wis-dom (respectively, semi-wisdom)? Are we willing and ready to operationalize it via judicious strategies for managing (or even controlling) the global environ-ment & development process (see Sect. 3)? And – what is the alternative to such a "geo-cybernetic" enterprise (Schellnhuber and Kropp 1998)?

When answering these nasty questions, we must be aware of two crucial facts: First, after the convulsions and catastrophes of the past century, which boasted again and again to create paradise on Earth through scientific-technological progress and ingenious large-scale planning (Schellnhuber 1998b), public long-term management has generally fallen into disrepute in the intellectual community. Second, "business as usual" is no less a geo-cybernetic strategy, which may transform the world faster and deeper than any sophisticated planning scheme. The so-called "globalization" as driven by the teleconnected forces of economy, technology, and life style is pushing forward the material development of many countries, yet modifies the environmental and social make-up of $\mathcal{E}$ beyond recognition – and, maybe, beyond stability. Thus we do not have the comfortable choice of keeping our place as it used to be: the remaining choice is between frantic, unbridled Global Change, and a co-evolution of nature and civilization which is kept on a tolerable long-term track by modest, careful and responsible intervention. The second alternative has become famous under the elusive label *Sustainable Development*.

Here we watch the grand entrance of the Global Subject $\mathcal{S}$, who will reign, without any doubt, over the centuries to come. In fact, $\mathcal{S}$ has been busy with geo-cybernetics for quite a while already, and the building and application of macroscopes (see sect. 2) have helped it to find its identity: An ever-evolving Earth-

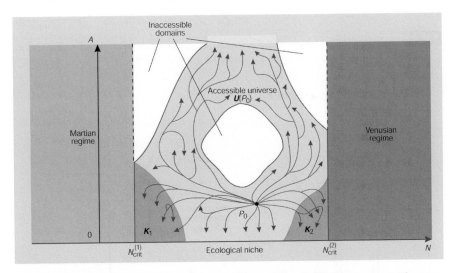

Fig. 18. A "theatre world" for representing paradigms of Sustainable Development. The space of all conceivable co-evolution states P = (N, A) is spanned by a "natural" axis, N (representing, say, global mean temperature) and a civilizatory axis, A (representing, say, global gross product). Vertical lines at $N_{crit}^{(1)}$ and $N_{crit}^{(2)}$ delimit the niche of subsistence states for humanity between an ultra-cold "Martian regime" and an ultra-hot "Venusian regime". The domain $U(P_0)$ ("accessible universe") embraces all possible co-evolution states that can be reached from the present state P_0 by some management sequence from the overall pool. $U(P_0)$ contains specific "catastrophe domains" K_1 and K_2.

observation system allow S to watch its own footprints on the ecosphere, Earth-simulation models enable S to plan interventions at the system's level, and densely linked global institutions as well as innumerable worldwide activists' network help enforce resolutions of S, such as those made in international environmental conventions. The Kyoto protocol (Oberthür & Ott 2000), for instance, is a portentous act of the immaterial planetary sovereign.

So the questions posed above boil down to a single one: What do we mean by Sustainable Development, i.e., what is the paradigm for the geo-cybernetic activities of S? The author has tried to derive the available options from first ethical and systemic principles, and to develop a rigorous formalism that allows to intercompare and implement those options (Schellnhuber 1998a). Some central results can be illustrated in a two-dimensional "theatre world", where Sustainable Development is played out as a strategic planning exercise (see fig. 18).

It should be emphasized that the overall co-evolution space includes two types of unpleasant domains: truly apocalyptic zones as exemplified by the Martian regime (hypothetically attainable through a run-away cooling-chamber process), or the Venusian regime (hypothetically attainable through a run-away greenhouse process) (Franck et al. 2000), and *catastrophe zones*, where humanity might subsist, but in a miserable manner. To be specific, we call some sub-set of co-evolution space a catastrophe domain, if this sub-set is (i) accessible from P_0, (ii) composed of strictly intolerable states for civilization (according to certain standards), and (iii) inescapable (by all means of geo-cybernetic management).

BOX 4

The five pure paradigms for Sustainable Development

Standardization:

Prescribing a long-term co-evolution corridor

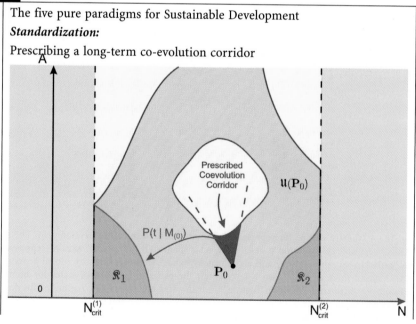

Optimization:

Maximizing an aggregate ecosphere – anthroposphere welfare function

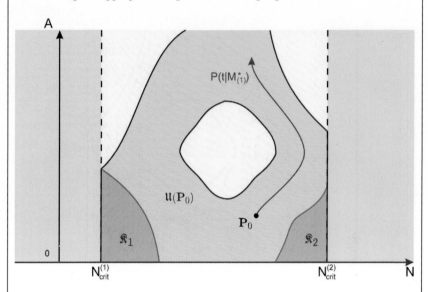

Pessimization:

Avoiding the worst under imperfect management

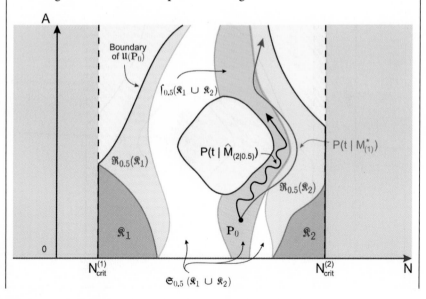

Equitization:

Preserving options for future generations

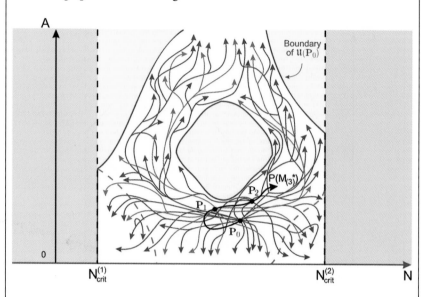

Stabilization:

Landing and maintaining the Earth System in a desirable state

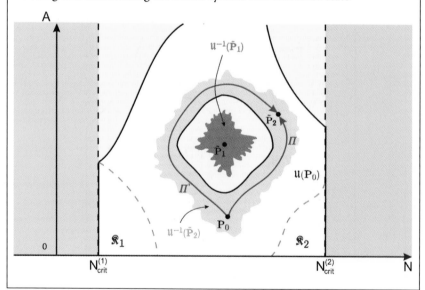

Five competing, "pure" paradigms for Sustainable Development can be identified on this systems-analytic basis by considering the principal human values. (1) *Standardization* – prescribing an explicit long-term co-evolution corridor emanating from P_0 in figure 18. (2) *Optimization* – getting the best, that is, maximizing an aggregated *$\mathcal{N}$-$\mathcal{A}$* welfare function by choosing the proper co-evolution segment over a fixed time period. (3) *Pessimization* – avoiding the worst, that is, steering well clear of catastrophe domains, allowing for the possibility of bad management by future decision-makers. (4) *Equitization* – preserving the options for future generations, not contracting the "accessible universe" over time. (5) *Stabilization* – bringing the *$\mathcal{N}$-$\mathcal{A}$* complex into a desirable (tolerable) state in co-evolution space and maintaining it there by good management. All these paradigms are again listed and caricatured in box 4 (for details, see Schellnhuber 1998a).

So, here is the menu from which the Global Subject can select its master principle, or suitable combinations thereof, for Earth-System control. The formal elaboration of these options and putting them into operation using, for example, EMICs, is a highly non-trivial exercise. Let us give here a few eclectic suggestions that may give an idea of the scope of the challenge. These suggestions are listed in descending order of speculativeness:

Optimization

The present geostrategic design of the *$\mathcal{N}$-$\mathcal{A}$* complex is far from being fair and effective. Evidently, the industrialized countries dominate most fields of global relevance like agricultural production, technological innovation, environmental protection and tourist services. This implies not only considerable inequity across the Earth's population, but also unsustainable distortions of the natural web of energetic and material fluxes. A global redesign could aim at establishing a more "organic" distribution of labour, where the temperate countries are the main producers of global food supplies, the sub-tropical zones generate renewable energies and high technology, and the tropical regions preserve biodiversity and offer a spectrum of recreation activities (like eco-tourism). This strategy may be supported by the responsible use of genetic engineering that will soon have all life on Earth at its disposal.

The entire concept may appear less utopian if we look back in time for a century and realize how much the world has changed since. After all, we can ameliorate our environmental resource basis much like the pioneers of the past who drained swamps and planted crops. For demonstrating that giant potentials still lie fallow, figure 19 gives an overview of those geographical locations on the planet which are both highly adequate for solar-energy harvesting and highly marginal with respect to other land-use types like agriculture or biodiversity conservation. As can easily be seen from this figure, even the locations in the highest category of suitability make up a significant amount of the terrestrial area. So, let us go solar ...

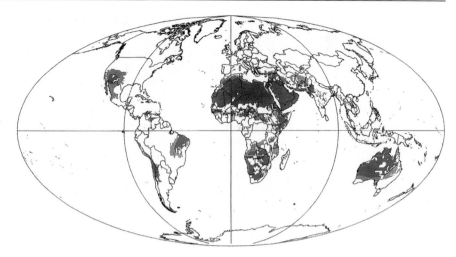

Fig. 19. Global distribution of sites available for the production of solarthermal and/or photo-voltaic energy. Suitability increases with intensity of trading.

Stabilization

Why should the Global Subject not try to mitigate the anthropogenic aberrations of the ecosphere – or even attempt to attenuate physiogenic excursions through natural variability that can have disastrous impacts on civilization? A number of relevant ideas have been put forward in the past years. The least sophisticated ones belong to the field of "geoengineering", where ecosphere repair schemes like iron fertilization of ocean regions for stimulating the marine "carbon pump" (Coale et al. 1996) or ozone-hole therapy through orbital laser surgery are proposed. But we can also think of more subtle stabilization strategies, in particular, schemes for "desensitizing" parts of the Earth's physique with respect to human and natural perturbations. One global fiction is concerned with the suppression of glaciations via "biome styling" (Schellnhuber and Kropp 1998) that thwarts, e.g., the dangerous taiga-tundra snow-albedo feedback (Otterman et al. 1984, de Noblet et al. 1996, Claussen 2000). This could be achieved by a portfolio of measures modifying the terrestrial vegetation, like the biotechnological hardening of boreal conifers against severe frosts. And the insights acquired during the present climate crisis may enable humanity to avoid future "ice ages" by judicious injection of "designer greenhouse gases" into the atmosphere.

Pessimization

Least speculative and most essential is the creation of a *manual of minimum safety standards* for operating the Earth System. Human interference with the ecosphere may provoke the perhaps irreversible transgression of critical thresholds, bringing about quantitatively different environmental conditions on a large scale. Among the singular events potentially accompanying unbridled Global

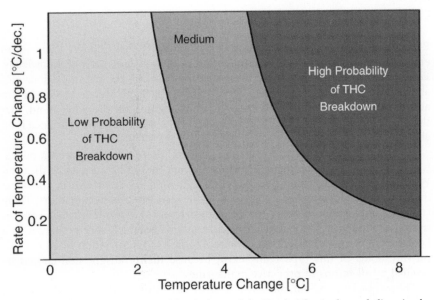

Fig. 20. Likelihood domains for total break-down of the North Atlantic thermohaline circulation.

Change, there are processes that even cannot be assigned low probabilities – the partial or total shut-down of ocean currents like the North Atlantic Drift, the destabilization of the West Antarctic and the Greenland ice sheet, or the abrupt die-back of tropical forests due to super-critical global warming. These "special effects" will constitute one of the foci of the forthcoming Third Assessment Report of the Intergovernmental Panel on Climate Change (IPCC).

The results of global research programmes like IGBP and WCRP will soon enable us to identify and respect appropriate "guardrails" for responsible planetary management that keep us away from the catastrophe domains (Petschel-Held et al. 1999a, Bruckner et al. 1999). These guardrails are, in general, rather "safe corridors" (Alcamo and Kreileman 1996), due to the irreducible cognitive deficits involved in the analysis of the crucial non-linear and complex processes. We demonstrate this for the possible suppression of North Atlantic Deep Water formation (generating, inter alia, the Gulf Stream) by anthropogenic perturbation of the planetary water cycle (see the reviews by Weaver and Hughes 1992, Rahmstorf et al. 1996, Rahmstorf 2000). Figure 20 is a "traffic-light diagram" that identifies the domains in global warming phase space, where a major interference with the ocean conveyor belt (Broecker 1987) is likely, possible, and unlikely, respectively (see also Stocker and Schmittner 1997).

The science of "guardrailing" (Schellnhuber and Yohe 1998) is still in its infancy, but it may eventually turn out to be the most robust tool for sustainability science in general.

Coping with Uncertainty

Recently, it has become quite fashionable among Global-Change scientists to emphasize the vagueness, limitedness and speculativeness of our knowledge. From the analytic point of view, these cognitive deficits are an almost unavoidable consequence of the matchless complexity of the Earth System as sketched in all the preceding sections. This state of affairs implies, on the other hand, that there is still so much to be unravelled in the field – and that is what science is all about...

From the normative (i.e., ethical and political) point of view, however, persistent ignorance about the planetary dynamics appears to be a veritable disaster: We are rapidly transforming our specimen, $\mathcal{E}$, *while we scrutinize it,* and most insights might come too late for correcting our ways towards sustainability. There is, for instance, some evidence for bifurcation levels of atmospheric CO_2 concentration that separate global futures with and without ice sheets (Berger and Loutre 1997, Berger 1999). It would be nice to find out rather soon whether critical thresholds of this type really exist, because their transgression should be avoided by all means of proactive management.

Thus knowledge about the Earth System must and will be turned into global action by responsible decision-makers. This means, in fact, that we are also transforming $\mathcal{E}$ *by scrutinizing it.* Although reminiscent of the quantum-mechanical observation problem (see, e.g., Claussen's contribution in this book), the situation is even more intricate here: For we can even scrutinize ourselves watching the virtual future and the repercussions thereof for the real future. In particular, all conceivable adaptation and mitigation strategies have to be taken into account. This self-referential process can be continued, in principle, *ad infinitum* by perpetually feeding back the simulated long-term consequences into the present-day decision-making. There are only two ways out of such a dilemma, namely (i) to truncate the loop at some arbitrary iterative level, or (ii) to achieve convergence towards a unique ("dominant") strategy for shaping the future by superimposing appropriate boundary conditions reflecting absolute principles ("categorical imperatives for sustainable co-evolution").

The real cognitive situation is, of course, much less complicated and will remain so over the next decades. As a matter of fact, our wisdom about Earth-System dynamics is still so primitive that genuine self-referentiality is out of reach. Thus our collective freedom of will, i.e. the long-term autonomy of the Global Subject, is firmly based upon deep ignorance (Schellnhuber 1998a). In the short run, however, it is the lack of knowledge that may force us on undesired, unpleasant co-evolution pathways that radically reduce the self-determination options of future generations. Yet this dangerous effect of scientific uncertainty can be confined to a minimum by maximum *responsiveness.*

The last statement clearly requires some justification. So let us recall the fact that global environmental management is not a prophecy contest but a cybernetic task. And it is often much easier to control a complex system than to predict its long-term behaviour. Think, for example, of the nasty exercise of entering a mega-city like London or Paris by car, and to steer the vehicle – without accident and in a finite period of time – to a given target location characterized by no more an information than the label "10 Downing Street" or "Sacré Coeur". Even if a precise road-map were available, there would be no way to determine in advance a sequence of control actions that would bring the car towards its goal: no computer programme in the world could predict and process all the mid-way contingencies that are caused, inter alia, by other drivers, cyclists, pedestrians, policemen, criminals, dogs, and all sorts of distractions respectively seductions. And yet we master this task, and even nastier ones, in every-day life!

The secret of our innumerable victories over theoretically intractable problems is the "fuzzy control principle" which reads as follows (Schellnhuber and Kropp 1998 and references therein): *Based on uncertain and/or fragmentary information, adopt a rough long-term and/or large-scale strategy, which is continuously readjusted in an approximate fashion in response to all sorts of generally imprecise additional information.*

Note that crisp data are not needed for this type of control, which fails, of course, in cases where no leeway for steering actions exists. In most situations, however, the lack of precision is easily compensated for by perpetual readjustment. It seems reasonable to assume that the Sustainable-Development challenge also defines such a situation. Thus the mantra for Earth-System managers should be:

Vision and Revision!

This means that the Global Subject takes an initial step in an appropriate direction in option space. The step is guided by scientific evidence from macroscopes like EMICs which provide, at least, hazy previews of possible co-evolution futures, and by the preferred paradigm for Sustainable Development. Whenever new scientific information arrives due to observation, theory and simulation modelling, or whenever new normative boundary conditions emerge due to modified preferences of S, the chosen direction may be corrected. The basic principle is simple, but its implementation may constitute a formidable task. Just think of the Kyoto Protocol and its associated cosmos of processes and institutions: its target period is 2008 – 2012, but there is little doubt that the general cognitive progress will have made most of its regulations obsolete by that time. So a major quality of global environmental conventions should be *flexibility*.

Here we conclude our brief *tour d'horizon* on Earth System Analysis and Management. It may have been a rough ride for many readers, who might either ask for more solid ground or better means of transportation. In fact, the following contributions by a number of eminent scientists will offer a wealth of well-exposed specific insights. Don't miss these signposts towards the Second Copernican Revolution!

Acknowledgements

I would like to thank my colleagues at the Potsdam Institute for intellectual and technical support during the preparation of this article. The kind assistance of A. Block, T. Bruckner, M. Cassel-Gintz, M. Claussen, W. Cramer, A. Ganopolski, H. Hoff, M. Lüdeke, I. Meyer S. Lütkemeier and S. Rahmstorf is particularly appreciated.

References

Alcamo J, Kreileman E (1996) Emission scenarios and global climate protection, Glob Environ Change 6: 305–334

Alcamo J, Leemans R, Kreileman E (1998) Global Change Scenarios of the 21st Century. Results from the IMAGE 2.1 Model. Pergamon, Oxford

Bechtel RB, MacCallum T, Poynter J (1997) Environmental Psychology and Biosphere 2. In: Wagner S et al. (eds) Handbook of Japan-United States Environment-Behavior Research: Towards a Transactional Approach. Plenum Press, New York

Berger A, Loutre MF (1997) Paleoclimate Sensitivity to CO_2 and Insolation. Ambio 26: 32–37

Berger A (1999) Personal Communication

Broecker W (1987) Unpleasant surprises in the greenhouse? Nature 328: 123–126

Bruckner T, Petschel-Held G, Toth FL, Füssel HM, Helm C, Leimbach M, Schellnhuber HJ (1999) Climate change decision-support and the tolerable windows approach. Environ Mod Assess 4: 217–234

Bunde A, Schellnhuber HJ (2001) The Science of Disaster: Clinical Disruptions, Hearth Attacks and Market Crashes. Springer, Berlin, in preparation

Cassel-Gintz M, Lüdeke MKB, Petschel-Held G, Reusswig F, Plöchl M, Lammel G, Schellnhuber HJ (1997) Fuzzy logic based global assessment of the marginality of agricultural land use. Clim Res 8: 135–150

Clark WC (1989) Managing planet Earth. Sci Am 261 (9): 46–54

Claussen M (2000) Biogeophysical feedbacks and the dynamics of climate. In: Schulze ED et al. (eds): Global Biogeochemical Cycles in the Climate System. Adacemic Press, San Diego, in the press

Claussen M, Cramer W (1999) Earth-System Models of Intermediate Complexity. Research GAIM Newsletter 3 (1): 8

Coale KH et al. (1996) A massive phytoplankton bloom induced by an ecosystem-scale iron fertilization experiment in the equatorial Pacific Ocean. Nature 383: 495–501

Defries R, Belward A (2000) Global and Regional Land Cover Characterization from Satellite. Int. J Remote Sensing 21: 1083–1092

de Noblet N, Prentice IC, Jousaume S, Texier D, Botta A, Haxeltine A (1996) Possible role of atmosphere-biosphere interactions in triggering the last glaciation. Geophys Res Lett 23: 3191–3194

de Rosnay J (1973) Le macroscope. Vers une vision globale. Seuil Paris

Dick SJ (1998) Life on other worlds. Cambridge University Press

Drake F, Sobel P (1992) Is anyone out there? The scientific search for extraterrestrial intelligence. Delacorte Press New York

Ducklow H (1999) Personal Communication

Franck S, von Bloh W, Bounama C, Steffen M, Schönberner D, Schellnhuber HJ (1999) Determination of habitable zones in extrasolar planetary systems: Where are Gaia's sisters? J Geophys Res 105: 1651–1658

Franck S, Block A, von Bloh W, Bounama C, Schellnhuber HJ, Svireshev Y (2000) Reduction of biosphere life span as a consequence of geodynamics. Tellus B 52: 94–107

German Advisory Council on Global Change (1995) World in Transition: The Threat to Soils. Economica, Bonn

German Advisory Council on Global Change (1997) World in Transition: The Research Challenge. Springer, Berlin

German Advisory Council on Global Change (2000) World in Transition: Conservation and Sustainable Use of the Biosphere. Earthscan, London, Berlin, in the press

Gordon C, Cooper C, Senior CA, Banks HT, Gregory JM, Johns TC, Mitchell JFB, Wood RA (2000) The simulation of SST, sea ice extents and ocean heat transports in a version of the Hadley Centre coupled model without flux adjustments. Clim Dyn 16: 147–168

GTOS (1998) GTOS Implementation Plan. GTOS document 17. ftp://ext-ftp.fao.org/GTOS/gtosplan.pdf

Hart MH (1979) Habitable zones about main sequence stars. Icarus 37: 351–357

IGBP (1999) Structure Viewgraphs Series by the International Secretariat, Stockholm, Item 5

Jolly A (1999) The fifth step. New Scientist 2218: 78–79

Justice C, Ahern F, Freise A (1999) Regional Networks for Implementation of the Global Observation of Forest Cover Projects in the Tropics. START Report No. 4, Proceedings of the Joint START/GOFC/CCNS Workshop, March 15–17

Kasting JF (1997) Habitable zones around low mass stars and the search for extraterrestrial life. Origins of Life 27: 291–307

Kibby H (1996) The Global Climate Observing System. Bulletin of the World Meteorological Organization 45, No. 2

King MD, Greenstone R (1999) 1999 EOS Reference Handbook. A Guide to NASA's Earth Science Enterprise and the Earth Observing System (NASA, Maryland). Also at http://eos.nasa.gov

Kleypas JA et al. (1999) Geochemical consequences of increased atmospheric carbon dioxide on coral reefs. Science 284: 118–120

Kuipers B (1994) Qualitative Reasoning: Modelling and Simulation with Incomplete Knowledge. MIT Press, Cambridge, Massachusetts

Lenton T (1998) Gaia and natural selection Nature 394: 439–447

Lovelock JE (1991) Gaia. The practical science of planetary medicine. Gaia, London

National Research Council (1999) Our Common Journey: a transition towards Sustainability. National Academy Press, Washington

Nordhaus WD (1992) An optimal transition path for controlling greenhouse gases. Science 258: 1315–1319

Oberthür S, Ott HE (eds) (2000) International climate policy for the 21st century. Springer, Berlin

Otterman J, Chou MD, Arking A (1984) Effects of nontropical forest cover on climate. J Clim Appl Met 23: 762–767

Petit JR et al. (1999) Climate and atmospheric history of the past 420,000 years from the Vostok ice core, Antarctica. Nature 399: 429–436

Petoukhov V et al. (2000) CLIMBER 2: a climate system model of intermediate complexity. Part I: Model description and performance for present climate. Clim Dyn 16: 1–17

Petschel-Held G, Schellnhuber HJ, Bruckner T, Toth FL, Hasselmann K (1999a) The tolerable windows approach: theoretical and methodological foundations. Clim Change 41: 303–331

Petschel-Held G, Block A, Cassel-Gintz M, Kropp J, Lüdeke MKB, Moldenhauer O, Reusswig F, Schellnhuber HJ (1999b) Syndromes of Global Change: a qualitative modelling approach to assist global environmental management. Environ Mod Assess 4: 295–314

Platt T, Sathyendranath S, Longhurst A (1995) Remote-sensing of primary production in the ocean – promise and fulfillment. Phil Trans Roy Soc B 348: 191–201

Pope VD, Gallani ML, Rowntree PR, Stratton RA (2000) The impact of new physical parameterisations in the Hadley Centre climate model. Clim Dyn 16: 123–146

Prentice IC, Cramer W, Harrison SP, Leemans R, Moserud RA, Solomon AM (1992) A global biome model based on plant physiology and dominance, soil properties and climate. J Biogeogr 19: 117–134

Prentice IC, Sykes MT, Lautenschlager M, Harrison SP, Denissenko O, Bartlein PJ (1993) Modelling global vegetation patterns and terrestrial carbon storage at the last glacial maximum. Glob Ecol Biogeogr Lett 3: 67–76

Rahmstorf S, Marotzke J, Willebrand J (1996) Stability of the thermohaline circulation. In: Krauss W (ed) The warm water sphere of the North Atlantic ocean. Borntraeger, Stuttgart, pp 129–158

Rahmstorf S (2000) The thermohaline ocean circulation: A system with dangerous thresholds? Clim Change in the press

Rotmans J, de Vries HJM (1997) Perspectives on global change: The TARGETS approach. Cambridge University Press

Running SW (1998) A Blueprint for Improved Global Change Monitoring of the Terrestrial Biosphere. The Earth Observer 10 (1), http://eospso.gsfc.nasa.gov/eos-observ/1-2-98/jan-feb-98.html

Saltzmann B (1978) A survey of statistical-dynamical models of the terrestrial climate. Adv Geophys 20: 183–304

Saltzmann B (1988) Modelling the slow climatic attractor. In: Schlesinger ME (ed) Physically-based modelling and simulation of climate and climatic change. Part II. Kluwer, Dordrecht, pp 737–754

Schellnhuber HJ (1998a) Earth System Analysis: The Scope of the Challenge. In: Schellnhuber HJ, Wenzel V (eds) Earth System Analysis. Integrating Science for Sustainability. Springer, Berlin, pp 3–195

Schellnhuber HJ (1998b) Globales Umweltmanagement oder Dr. Lovelock übernimmt Dr. Frankensteins Praxis. In: Altner G et al. (eds) Jahrbuch Ökologie 1999. C.H. Beck, München, pp 168–186

Schellnhuber HJ (1999) 'Earth system' analysis and the second Copernican revolution. Nature 402, Supp. 2 Dec 1999: C19–C23

Schellnhuber HJ, Kropp J (1998) Geocybernetics: Controlling a Complex Dynamical System under Uncertainty. Naturwissenschaften 85: 411–425

Schellnhuber HJ, Yohe G (1998) Comprehending the Economic and Social Dimensions of Climate Change by Integrated Assessment, in Achievements, Benefits and Challenges. Proceedings of the WCRP Conference, 26–28 August 1997, Geneva. WMO, Geneva, pp 179–198

Schellnhuber HJ, Block A, Cassel-Gintz M, Kropp J, Lammel G, Lass W, Lienenkamp R, Loose C, Lüdeke MKB, Moldenhauer O, Petschel-Held G, Plöchl M, Reusswig F (1997) Syndromes of global change. GAIA 6: 19–34

Schellnhuber HJ et al. (2000) Syndromes & Company: Semiquantitative Approaches in Global Change Research. In Proc. Conf. "Transsektorale Forschung zum Globalen Wandel", Bonn, 27–28 January 2000, forthcoming

Schneider J (1999) The Extrasolar Planets Encyclopaedia. http://www.usr.obspm.fr/planets

Schneider SH (1997) Integrated assessment modeling of global climate change: Transparent rational tool for policy making or opaque screen hiding value-laden assumptions? Environ Mod Assess 2: 229–249

Schneider SH, Boston PJ (1993) Scientists on Gaia. MIT Press, Cambridge, Massachusetts

Sellers P et al. (1995) Remote Sensing of the Land Surface for Studies of Global Change: Models – Algorithms – Experiments. Remote Sens Environ 51: 3–26

Stocker TF, Schmittner A (1997) Influence of CO_2 emission rates on the stability of the thermohaline circulation. Nature 388: 862–865

Stouffer RJ, Tett SFB, Hegerl G (1999) A comparison of surface air temperature variability in three 1000-year coupled ocean-atmosphere model integrations. J Climate 13: 513–537

Takahashi T et al. (1997) Global air-sea flux of CO_2: An estimate based on measurements of sea-air CO_2 difference. Proc Natl Acad Sci USA 94: 8292–8299

The Energy Journal (1999) Special Issue – The Costs of the Kyoto protocol: A Multi-Model Evaluation

Timmermann A et al. (1999) Increased El Niño frequency in a climate model forced by future greenhouse warming. Nature 398: 694–697

Volk T (1998) Gaia's Body. Copernicus, New York

Voss R, Mikolajewicz U (2000) Long-term climate changes due to increased CO_2 concentration in the coupled atmosphere-ocean general circulation model ECHAM3/LSG. Clim Dyn, accepted

Walker DA (1997) Advances in General Equilibrium Theory. Edward Elgar, Cheltenham

Walker B, Steffen W, Canadell J, Ingram J (1999) The Terrestrial Biosphere and Global Change. Cambridge University Press

Walter U (1999) Zivilisationen im All: Sind wir allein im Universum? Spektrum, Heidelberg

Weaver AJ, Hughes TMC (1992) Stability and Variability of the thermohaline circulation and its link to climate In: Council of Scientific Research Integration: Trends in physical oceanography. Trivandrum, India, pp 15–70

Wright DH (1990) Human impacts on energy flow through natural ecosystems and implications for species endangerment. Ambio 19: 189–194

CHAPTER 3

The Earth System: A Physiological Perspective   3

Peter S. Liss*

A little over thirty years ago mankind had the first opportunity to view planet Earth directly from the Moon. It can be argued that this represented a major turning point in how we have regarded our home planet ever since. Seen from the Moon, several aspects of the Earth, each of which was known in an intellectual way previously, became strongly imprinted on our consciousnesses. Viewing Earth from the Moon strongly reinforced the idea that our planet must be seen and studied as a whole, and not split into component parts, as has been done previously in most academic research. Further, the isolation of the earth in the vastness of space was made abundantly clear. In addition, the obvious dominance of the oceans in terms of coverage relative to land led to the idea that the Earth should really have been called 'Ocean'. Finally, the blue oceans, green/brown land, and white clouds all looked very different from the colouring of the other planets we can see using telescopes from Earth.

This last aspect is due in large measure to the existence of living organisms on the Earth. In this chapter some of the ways in which the functioning (hence the word 'physiology' in the title) of the Earth's biota controls important properties of the atmosphere, oceans and land are discussed. For the atmosphere, comparison of its chemical composition with that of other planets strongly shows the influence of biological processes. Many reduced gases exist at concentrations far in excess of what purely inorganic processes would allow. In the oceans the biota play a strong role in fixing the amounts of gases important in controlling the Earth's radiation balance, such as carbon dioxide and dimethyl sulphide. It has recently been shown how delicately poised the oceanic biota are in terms of iron availability, and hence in their ability to control levels of these and other gases in the atmosphere. The land plants are efficient exchangers of carbon dioxide and water vapour with the atmosphere. They control the seasonal cycle of the former in the atmosphere and their response in terms of uptake/release of water and carbon dioxide through their stomata may play a large role in how the Earth's climate responds to future rising levels of greenhouse gases.

* e-mail: p.liss@uea.ac.uk

The Composition of the Atmosphere

The chemical composition of the Earth's atmosphere clearly shows the influence of biospheric processes on land and in the oceans. A graphic way of illustrating this is given in figure 1, which shows the composition of the atmosphere with life present (e.g. as now) and a prediction of what it would be like in the absence of life.

Such a prediction of the lifeless state is made by assuming the Earth and its atmosphere to be at thermodynamic equilibrium. In contrast, the presence of life puts it out of equilibrium. It should be noted that in figure 1 the percentage contribution of each gas to the total composition is shown as a histogram on a logarithmic scale. So, for example, the change in carbon dioxide between the thermodynamic (dead) and living states is four orders of magnitude. In this chapter I use the figure as a framework to discuss some of the ways in which land and ocean biological activity can alter atmospheric composition. The intent is not to be comprehensive (which is not possible in a short article) but to give several illustrations of how life contributes to the non-equilibrium state. I begin with a very simple consideration of the thermodynamic basis of life on our planet and follow with brief examples of how biological processes affect atmospheric composition for gaseous forms of the three elements carbon, sulphur and nitrogen.

Fig. 1. Life's influence on the composition of the Earth's atmosphere. Percentage composition plotted on a logarithmic scale for the atmosphere with and without life present on the planet.
Source: IGBP (1992), after Margulis L and Lovelock JE (1989).

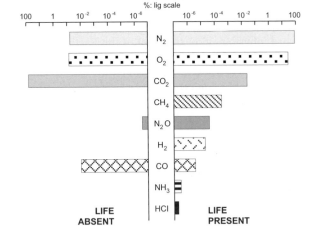

The Thermodynamics of Life on Earth

The fundamental equation describing biological activity at the Earth's surface is as follows:

$$\text{photosynthesis}$$
$$(h\mu, N, P, Fe, ..)$$
$$6CO_2 + 6H_2O \Leftrightarrow C_6H_{12}O_6 + 6O_2 \quad \Delta G = +2879 \text{ kJ}$$
$$\text{respiration/}$$
$$\text{decomposition}$$

The forward reaction represents the process of photosynthesis in which carbon dioxide combines with water in the hugely complex set of reactions, which occur in plants, to form carbohydrate material (here represented by the formula $C_6H_{12}O_6$). In the process of photosynthesis oxygen is released. This formally simple equation is the major explanation for the decrease in carbon dioxide and increase in oxygen between the life present and life absent cases shown in figure 1. The advent of photosynthesising organisms on Earth led to a drawdown of CO_2 and concomitant release of O_2, with a fraction of the synthesised carbohydrate material being buried in ocean sediments and in soils and other carbon stores on land. As well as CO_2 and H_2O, photosynthesis also requires other reactants including nitrogen, phosphorus and other nutrients including iron.

Also shown in the equation is the Gibbs Free Energy (ΔG) for the reaction. For photosynthesis the sign of ΔG is positive which means that the reaction will not occur spontaneously and a source of energy is required in order for it to take place. Of course, on Earth this energy (denoted by $h\mu$) comes from the sun, and 2879 kJ of solar energy are needed in order to produce one mole of carbohydrate. The reverse reaction, in which plant material is 'burnt' in atmospheric oxygen to give CO_2 and H_2O, has a negative sign for ΔG which indicates that it will occur spontaneously with release of 2879 kJ of energy for every mole of carbohydrate consumed. The back reaction and its ability to supply energy is the basis of decomposition and respiration processes, on which for example bacteria and humans, respectively, rely for their existence.

Carbon in the Form of Carbon Dioxide (CO_2)

The photosynthesis-respiration/decomposition equation discussed previously is not only an important driver in lowering atmospheric concentrations of carbon dioxide on our planet compared with a lifeless one, but can also help us understand the seasonal cycles of CO_2 and oxygen in the atmosphere. In figure 2 the observational record of atmospheric CO_2 at Mauna Loa Observatory on Hawaii from 1958 is shown, together with the somewhat less complete set of measurements from the South Pole.

The trend of increasing atmospheric CO_2 due to anthropogenic burning of coal, gas and oil (which can be viewed as a special case of the respiration reaction in our thermodynamic equation, but with hydrocarbon rather than carbohydrate

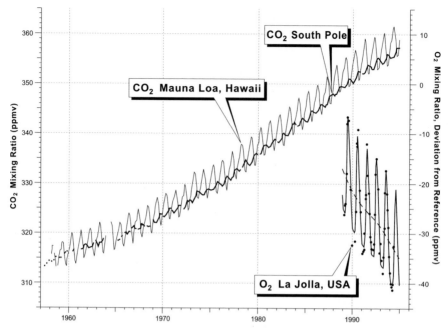

Fig. 2. Direct measurements of the atmospheric CO_2 concentration at Mauna Loa (Hawaii) and the South Pole, together with measurements of O_2 in the atmosphere at La Jolla.
Source: Heimann (1998).

as the 'fuel') is clear at both measurement sites. What is more interesting in the present context is the saw-tooth annual cycle which is superimposed on the general upward trend. This is due to removal of CO_2 in the spring/summer seasons due to photosynthesis and its return to the atmosphere in the autumn and winter when respiration and decomposition are the dominant operative processes. The effect is large at Mauna Loa due to its location in a latitude band where the land biology is extensive and seasonal, which contrasts with the small amplitude of the seasonal signal at the South Pole due to the lack of biological activity. The two seasonal cycles are 6 months out of phase, as is to be expected given that the locations are in opposite hemispheres. Since the late 1980s it has been possible to monitor the atmospheric concentration of oxygen, and this record is also shown in figure 2. As is predictable from the thermodynamic equation, oxygen concentrations are falling due to its consumption in the burning of fossil fuels, and the seasonal cycle is in anti-phase with that for CO_2. Thus, although the equation is an extremely simplified description of the reality of Earth surface biology, it does capture in a quantitative way the essential biogeochemistry and the role it plays in controlling the concentrations of vital atmospheric gases.

The almost 40 year length of the Mauna Loa CO_2 record has enabled Keeling et al. (1996) to use it to assess whether the amplitude of the biological 'pumping', which the seasonal cycle clearly shows, has changed over that period. Figure 3

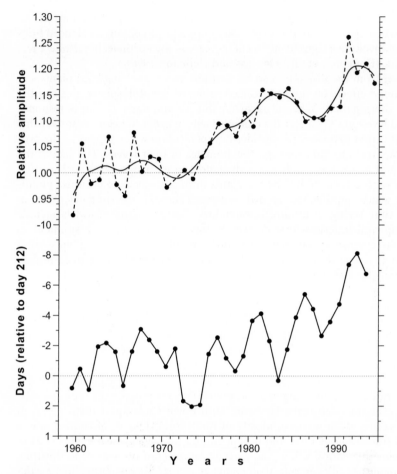

Fig. 3. Trends in the relative amplitude and timing of the seasonal cycle of atmospheric CO_2 at Mauna Loa.
Source: Keeling et al. (1996).

gives both the relative amplitude and the timing of the seasonal cycle, as deduced from the Mauna Loa record, as a function of time.

It is clear that since the mid–1970s the amplitude of the seasonal signal has increased by approximately 20 % and the growing season has lengthened by about 7 days. These results indicate an intensification of the annual cycling (but not necessarily carbon storage) of CO_2 between the northern hemisphere land biosphere and the atmosphere. There are three potential drivers of this effect: increasing levels of carbon dioxide in the atmosphere, enhanced inputs of plant nutrients or increasing global temperatures. All three are likely to result in enhanced uptake and release of CO_2 by land plants, but it is not currently possible to assess their relative importance. Whatever the mechanism, the effect is important in the context of trying to account for the anthropogenic part of the current carbon cycle, where it is essentially impossible to obtain a balanced budget without invoking a significant recent increase in the northern hemisphere terrestrial sink (Houghton et al. 1996, Liss and Heimann 1998).

It is clear from the discussion so far that processes involving land plants can have profound effects on the CO_2 concentration of the atmosphere. But can the plants, by altering their exchange of carbon dioxide and water vapour in response to elevated levels of CO_2, affect the climate itself? In order to begin to answer this question it is first necessary to consider how plants lose water vapour in the process of evapotranspiration through the stomates of their leaves. The size of the stomates and hence their ability (conductance) to exchange gases such as CO_2 and water vapour is controlled by the plants in such a way as to optimise photosynthesis, which requires CO_2 to diffuse inward through the stomates, while at the same time trying to minimise water loss. Current climate models include simple empirical functions that determine the conductance of the stomates in terms of the environmental state. In a world of enhanced CO_2 the plants could operate so as to use water more efficiently, i.e. they would be able to assimilate more carbon for the same amount of water loss. It is expected that this effect would only be important in areas with substantial dry periods. But what if the plants are able to adapt to the higher CO_2 concentrations by simply developing less stomates (called downregulation)? This would mean that the higher CO_2 concentration would lead to a large reduction in the conductance of plant stomates everywhere. To see whether these processes can have an effect globally, Sellers et al. (1996) ran a general circulation climate model with and without the CO_2-induced downregulation of the stomatal density of plants. They found that without downregulation a doubling of atmospheric CO_2 resulting in the typical 2–3 °C warming over the continents. However, with downregulation a further 1 °C of warming occurs at the surface, mostly in the tropics. This is, of course, only a computer simulation of what might happen and we have no real evidence as to whether downregulation will actually occur. However, it does serve to illustrate the potentially significant role that plants may play in regulating the Earth's future (and past) climate.

So far I have largely concentrated on the role of the land plants in affecting atmospheric CO_2 and hence climate. However, the oceans also have a large potential for altering the composition of the atmosphere through uptake and release of CO_2 and other gases which can affect the Earth's radiation balance. The oceans take up and release CO_2 either by physical dissolution and chemical reaction (the physical pump) or by the biological processes described in the thermodynamic equation (the biological pump). A frequently asked question is whether in the past or potentially in the future the biological pump might take up/release more or less CO_2.

Modelling studies certainly indicate that the marine biosphere has the potential to bring about large changes in atmospheric CO_2 levels. For example, it has been calculated that for an ocean devoid of biological life the CO_2 concentration in the atmosphere would be 450–460 ppm (Shaffer 1993). On the other hand, if the marine biota were to achieve their maximum potential productivity, modelling by Sarmiento and Toggweiler (1984) indicates that atmospheric CO_2 could be drawn down to 165–185 ppm. However, these are extreme cases, although they do illustrate the potential importance of changes in marine biological activity.

There is no firm evidence that the biological pump has altered due to man-induced global changes since industrialisation. However, it is likely that on longer time scales there have been natural changes in marine productivity. For example, several lines of evidence point to higher marine production during the last glaciation (Berger et al. 1989), which would have contributed to the lower atmospheric CO_2 levels at that time (see fig. 5).

In modelling studies of how changes in both the physical and biological pumps might be affected in a global warming scenario due to rising atmospheric CO_2 concentrations, Maier-Reimer et al. (1996) and Sarmiento and Le Quéré (1996) show that both oceanic pumps are potentially of importance. They find that the predicted global warming leads to a diminution of the physical pump due to a weakening of the oceanic thermohaline circulation, so that less CO_2 is taken up. However, this is compensated, at least in part, by an increase in the CO_2 taken up by the biological pump producing an enhanced downwards flux of carbon. However, our knowledge of how the oceans in general and the marine biota in particular will respond to likely future scenarios is still very rudimentary. Topics concerning the biological pump requiring further study include: the role of circulation changes in supplying nutrients to the surface waters; the potential for changes in iron inputs from the atmosphere to affect marine productivity (see later) and carbon removal to deep waters and its ultimate burial; the importance of increased atmospheric nitrogen inputs from pollution sources and from possible changes in nitrogen fixation in the surface oceans; the effect of increases in surface water acidity due to doubling (–0.25 pH units) and quadrupling (–0.5 pH units) of atmospheric CO_2.

Sulphur in the Form of Dimethyl Sulphide (DMS)

Another important example of how biological activity can have a profound effect on the atmosphere, in this case again involving the marine biota, is the gas dimethyl sulphide. DMS is formed in the ocean from a precursor compound (dimethylsulphoniopropionate – DMSP) which marine phytoplankton make as an osmolyte to help them cope with the high salt content of seawater. DMSP and DMS are released from the plankton to seawater, often following grazing by zooplankton or viral attack, where they are involved in a complex web of interactions (Liss et al. 1997). However, there is almost always a residual amount of DMS found in the surface water and because atmospheric levels are very low a net one-way flux to the atmosphere occurs. Once in the air DMS is unstable and subject to oxidation by free radical species to form variety of products, including sulphur dioxide, methanesulphonic acid (MSA) and sulphate particles. These oxidation products are generally acidic and it is this cycle which provides the natural acidity of marine rain and aerosol particles. Another property of the sulphate particles is to act as cloud condensation nuclei (CCN) in the marine atmosphere distant from land. The number density of CCN is an important parameter controlling the extent and type of cloud cover (and hence cloud albedo) over the oceans,

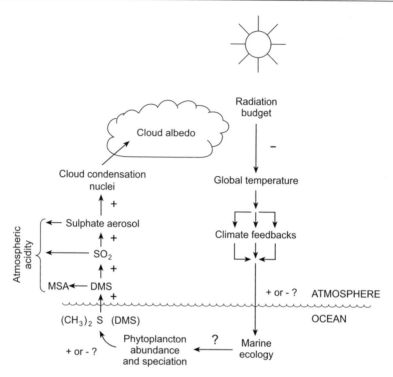

Fig. 4. Proposed feedback cycle between climate and marine DMS production. The + and –
signs indicate whether an increase in the value of the preceding parameter in the cycle is
expected to lead to an increase (+) or a decrease (-) in the value of the subsequent parameter.
Source: Andreae (1990).

and so plays a key part in determining the radiation balance of the Earth. These
roles of DMS oxidation products in the atmosphere are summarised in figure 4.

As indicated in figure 4, it has additionally been proposed that the DMS/CCN/
Cloud Albedo scheme may constitute a mechanism by which the planet's climate
is regulated (Charlson et al. 1987). For example, it is postulated that a rise in
atmospheric and surface ocean temperatures would produce an increase in DMS
production and its release to the atmosphere. This would result in an increase in
CCN number density and cloud albedo, which would tend to lower the temperature, and so counteract the original perturbation. If correct, this is an extremely
powerful idea, since the existence of such a negative feedback loop would mean
that biological processes would have a key role in controlling not only the chemistry of the atmosphere but also its physics, in stabilising the temperature. It has
to be said that the 'jury is still out' concerning the correctness and applicability
of the idea. Results from an Antarctic ice core covering the last ice age imply that
the colder ocean temperatures at that time led to an elevation in MSA (an oxidation product of DMS and a surrogate for it), relative, to present levels (see fig. 5),
which implies a DMS/Climate feedback of the 'wrong' sign. However, the complex-

ity of the system makes it very difficult to find a simple test for such a hypothesis, without interference from the myriad of other processes operating.

It should be pointed out that what has been discussed above is concerned with what happens in the Earth system unaffected by human activities. During the last century over a substantial part of the heavily industrialised/urbanised northern hemisphere the natural cycle has been very significantly supplemented with extra sulphur (emitted in the form of SO_2 – one of the major oxidation products of the natural source gas DMS) from the burning of sulphur-rich fossil fuels.

As mentioned earlier, the importance of iron as a control on oceanic primary production has recently been demonstrated in a series of in situ fertilisation experiments carried out in the equatorial Pacific and Southern Oceans (Martin et

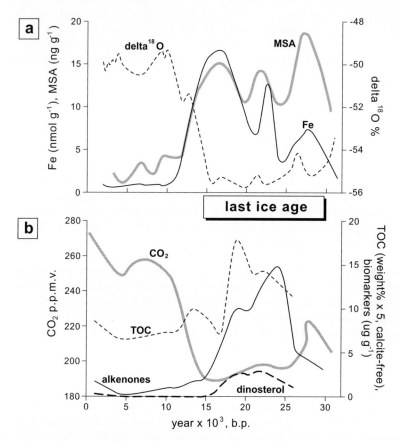

Fig. 5. Ice core and sediment data for the Holocene and end of the last ice age. a. Methane sulphonic acid (MSA) and $d^{18}O$ in an ice core from Dome C, east Antarctica; estimated Fe concentration in an ice core from Vostok, Antarctica. b. CO_2 concentration in an ice core from Vostok, Antarctica; total organic carbon (TOC), alkenones and dinosterol in a sediment core from the eastern tropical Pacific.
Source: Turner et al. (1996).

al. 1994, Coale et al. 1996, Boyd et al. 2000). In these studies DMS and its precursor DMSP were measured and it was found that significantly enhanced levels of both compounds resulted from additions of iron (the increase in biological activity also led to a substantial drawdown in CO_2). Not unexpectedly, the DMSP generally appeared before its breakdown product DMS. The global implications for climate of these results obtained with 100 km^2 scale patches are hard to assess. One way is to examine ice cores records covering the last ice age to the present. One such record is shown in figure 5.

The results in figure 5 show that in the last ice age atmospheric CO_2 was lower, and MSA (marker for DMS) and iron were both higher. The most obvious interpretation of this data is that ocean primary production was greater in the glacial ocean than now, with concomitant drawdown of atmospheric CO_2 into the ocean and increased emissions of DMS to the atmosphere. Enhanced production is supported by data from deep sea cores showing higher biomarker and total organic carbon levels during glacial times (see fig. 5). The fact that iron levels were also higher is a necessary condition, but not a proof (which correlation alone cannot provide), for establishing the importance of iron as a control on oceanic primary production.

Nitrogen in the Form of Ammonia (NH_3)

Ammonia is another example of a gas found in the atmosphere which would not occur there but for biological activity. In figure 1 the lifeless Earth's atmosphere shows essentially no NH_3, whereas with life present there is a small but significant amount. The ammonia comes from a variety of sources on land as well as from the oceans; here we will consider only the latter source because it can be treated as a companion air-sea flux to that of DMS discussed in the previous section. As in the case of sulphur, the nitrogen cycle has also been substantially amended recently through a variety of human activities.

In order to examine the role and behaviour of NH_3 in its natural setting now we have to go to parts of the globe furthest from man's activities. Since such areas are also far from land, this enables the marine part of the cycle to be examined without major influence from terrestrial source regions (this is possible because the atmospheric lifetime of NH_3 after emission from land (or sea) is only a matter of days). Figure 6 shows measurements made on atmospheric aerosol particles collected over a 5-year period at a coastal sampling site in Antarctica.

A clear seasonal cycle is evident for ammonium (the form of ammonia in the aerosol), with highest values in the spring and summer seasons and minimum levels in autumn and winter. Also remarkable in figure 6 is the similarity of the ammonium to the MSA and non sea-salt sulphate (both oxidation product of DMS) seasonal cycles. These similarities strongly suggest a marine biological origin for the NH_3, as well as for the DMS products. Support for this idea comes from a more detailed analysis of the data in figure 6, from which it can be calculated that the molar ratio of ammonium to non sea-salt sulphate in the aerosols is close to 1:1, implying a chemical composition of ammonium bisulphate (NH_4HSO_4). The final piece of the jigsaw comes from measurements made in the

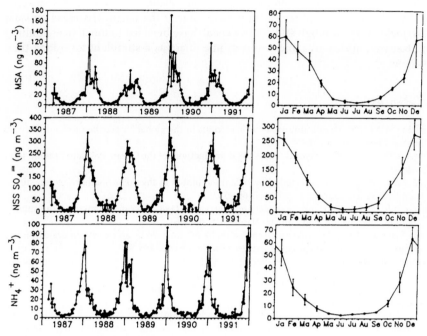

Fig. 6. Graphs on the left show the actual time series of the weekly-average concentrations of methanesulphonic acid (MSA), non sea-salt sulphate (NSS SO_4^{2-}) and ammonium (NH_4^+) at Mawson, Antarctica. Graphs on the right illustrate the composed seasonal cycles, showing the monthly means and their 95 % confidence intervals.
Source: Savoie et al. (1993).

north and south Pacific Ocean by Quinn et al. (1990), from which they were able to calculate the fluxes of both NH_3 and DMS being emitted from the ocean to the atmosphere. Although both fluxes show quite a lot of spatial variability, the mean values are again reasonably close to a 1:1 ratio, supporting the idea that the aerosol composition in terms of both nitrogen and sulphur (the main constituents of the total mass) can be explained by marine biogenic emissions.

Conclusions

My intention in this short chapter has not been to give a comprehensive account of all the ways the land and marine biota affect the composition of our atmosphere. Instead, I hope that enough has been said to show that at least for the major elements carbon, sulphur and nitrogen a strong interaction certainly exists, with important implications for the physiology of the planet. Much of what has been discussed here has been inspired by the pioneering work of Prof. Jim Lovelock; he was the first to make measurements of DMS in seawater (Lovelock et al. 1972) and has been instrumental in developing many of our concepts on the role of the biosphere in the functioning of the Earth. His ideas on Gaia (Lovelock 1979), although still controversial, have been led to much new thinking and research in this area and will continue to act as a stimulant for a long time to come.

References

Andreae MO (1990) Ocean-atmosphere interactions in the global biogeochemical sulfur cycle. Marine Chemistry 30: 1-29

Berger WH, Smetacek V, Wefer G (eds) (1989) Productivity of the Ocean: Past and Present. John Wiley

Boyd PW et al. (2000) A mesoscale Phytoplankton bloom in the polar Southern Ocean stimulated by iron fertilization. Nature 407: 695-702

Charlson RJ, Lovelock JE, Andreae MO and Warren SG (1987) Oceanic plankton, atmospheric sulphur, cloud albedo, and climate. Nature 326: 655-661

Coale KH et al. (1996) A massive phytoplankton bloom induced by an ecosystem-scale iron fertilization experiment in the equatorial Pacific. Nature 383: 495-501

Heimann M (1998) A review of the contemporary global carbon cycle and as seen a century ago by Arrhenius and Högbom. In: Rodhe H, Charlson R (eds) The Legacy of Svante Arrhenius Understanding the Greenhouse Effect. Royal Swedish Academy of Sciences and Stockholm University, pp 43-58

Houghton JT, Meira Filho LG, Callander BA, Harris N, Kattenberg A, Maskell K. (eds) (1996) Climate Change 1995. The Science of Climate Change, Contribution of Working Group I to the Second Assessment Report of the Intergovernmental Panel on Climate Change. Cambridge University Press, Cambridge

IGBP (1992) Global Change: Reducing Uncertainties. International Geosphere-Biosphere Programme

Keeling CD, Chin JFS, Whorf TP (1996) Increased activity of northern vegetation inferred from atmospheric CO_2 measurements. Nature 382: 146-149

Liss PS, Heimann M (1998) The biosphere and climate. In: Proceedings of the Conference on the World Climate Research Programme: Achievements, Benefits and Challenges, Geneva 26-28 August 1997, pp. 139-147

Liss PS, Hatton AD, Malin G, Nightingale PD, Turner SM (1997) Marine sulphur emissions. Phil Trans RSoc Lond B 352: 159-169

Lovelock JE, Maggs RJ, Rasmussen RA (1972) Atmospheric dimethyl sulphide and the natural sulphur cycle. Nature 237: 452-453

Lovelock JE (1979) Gaia: A New Look at Life on Earth. Oxford University Press, 157 pp

Maier-Reimer E, Mikolajewicz U, Winguth A (1996) Future ocean uptake of CO_2: Interaction between ocean circulation and biology. Climate Dynamics 12: 711-721

Martin JH et al. (1994) Testing the iron hypothesis in ecosystems of the equatorial Pacific Ocean. Nature 371: 123-129

Quinn PK, Bates TS, Johnson JE, Covert DS, Charlson RJ (1990) Interactions between the sulfur and reduced nitrogen cycles over the Central Pacific Ocean. Journal of Geophysical Research 95: 16, 405-16, 416

Sarmiento JL, Toggweiler JR (1984) A new model for the role of the oceans in determining atmospheric CO_2. Nature 308: 621-624

Sarmiento JL, Le Quéré C (1996) Oceanic carbon dioxide uptake in a model of century-scale global warming. Science 274: 1346-1350

Savoie DL, Prospero JM, Larsen RJ, Huang F, Izaguirre MA, Huang T, Snowdon TH, Custals L, Sanderson CG (1993) Nitrogen and sulfur species in Antarctic aerosols at Mawson, Palmer Station, and Marsh (King George Island). Journal of Atmospheric Chemistry 17: 95-122

Sellers PJ, Bounoua L, Collatz GJ, Randall DA, Dazlich DA, Loss SO, Berry JA, Fung I, Tucker CJ, Field CB, Jensen TG (1996) Comparison of radiative and physiological effects of doubled atmospheric CO_2 on climate. Science 271: 1402-1406

Shaffer G (1993) Effects of the marine biota on global carbon cycling. In: Heimann M (ed) The Global Carbon Cycle. Springer, pp 431-455

Turner SM, Nightingale PD, Spokes LJ, Liddicoat MI, Liss PS (1996) Increased dimethyl sulphide concentrations in sea water from in situ iron enrichment. Nature 383: 513-517

CHAPTER 4

The Earth System: An Anthropogenic Perspective 4

Rüdiger Wolfrum*

Environmental treaty law has developed most progressively since the beginning of 1970 (Sir Palmer 1992, Kiss 1983, Brown Weiss 1992/93a, pp 675 et seq., Jacobson and Brown Weiss 1995, pp 119, Werksman 1995, pp 30 et seq.). The development of this new body of law is due to the fact that the industrial development, the growth of population, and especially the necessity to safeguard a sustainable development[1] for all States and for future generations[2] requires States to co-operate more intensively. In particular, the development of new institutional forms of co-operation for the protection of the integrity and sustainability of the environment is necessary.[3]

International law does not provide for a generally accepted definition of environmental law (Kiss and Shelton 1991, Sands 1995b, Birnie and Boyle 1992). Earlier international environmental agreements concentrated on the protection, preservation or management of components of the natural environment with a view to providing for a sustainable exploitation or use. More recent international agreements are broader in scope, including, amongst others, also the protection of the intrinsic value of the environment. This element has to be included in to the term environment as used in principle 2 of the Stockholm Declaration on the Human Environment (Kiss and Shelton 1991).[4]

Not only the number of international environmental agreements, but also the scope they cover has expanded significantly. Existing international environmen-

---

* e-mail: rwolfrum@mpiv-hd.mpg.de

[1] Principle 27 of the Rio Declaration on Environment and Development commits States and people to further develop "international law in the field of sustainable development". The term "sustainable development" has been particularly emphasized by the Brundtland Report of the World Commission on Environment and Development although its underlying rationale has been implicitly recognized in earlier international instruments concerning the utilization of natural resources. As to this approach see Sands (1995a), Handl (1995), Beyerlin (1996), Malanczuk (1995).

[2] This aspect has been particularly emphasized by Brown Weiss (1990), Brown Weiss (1992/93b).

[3] See Rio Declaration on Environment and Development of 14 June 1992, Report of the United Nations Conference on Environment and Development (A/CONF.151/Rev.1), Vol. I, 3 which in its Preamble emphasizes the integral and interdependent nature of Earth and the need to establish a new and equitable global partnership. The need for an intensified co-operation has been underlined in several respects. See also Keohane et al. (1994).

[4] UN Doc. A/CONF.48/14/Rev. 1, ILM (1972) 11: 1416.

tal law contains, amongst others, regulations in respect of transfrontier damage to components of the environment and concerning the protection of those resources which are commonly shared by several States (rivers, lakes, living resources such as fish stocks). It also provides for the protection and management of wildlife, the protection and management of areas beyond national jurisdiction such as the high seas, outer space, and Antarctica, as well as the management of environmental problems of global relevance. This latter category embraces measures for the protection of the ozone layer, biodiversity and against climate changes; this list is not exhaustive.

The obligations imposed by the respective agreements – be it an obligation to take certain precautionary actions,[5] to adopt a particular legislation, to take administrative measures or the obligation to take or refrain from particular activities – have and will result in directing activities of States concerning the environment or components thereof and thus constitute restrictions to the sovereignty of the individual State (Brown Weiss 1992/93a, pp 686 et seq.). The establishment of these obligations, in effect, shifts the task to preserve and manage certain components of the environment from the individual States' responsibility to that of the entire State community.

International law may either order, prohibit or permit a particular content of national law as of its substance or only prescribe a specific procedure or a particular policy. Both, orders and prohibitions, imply per definitionem some kind of restraint on State activities. Norms of conduct tell what a subject, be it a State or an individual, must do or must refrain from doing. In national environmental law, norms of conduct appear as prohibitions of certain harmful activities, prescriptions on the use of best available technology, fixed emission standards and various principles on due diligence or precaution. In international environmental law, too, norms of conduct are defined in terms of complete or partial prohibitions, prescriptions on the establishment of particular legal procedures or obligations to adopt certain policies. These norms define the obligations of States and are of main concern when elaborating on the compatibility of international and national environmental law.

The International Law Commission in its draft articles on State responsibility distinguishes between two kinds of international obligations referred to as obligations of conduct and obligations of result.[6] The former is defined as an obligation which must be achieved through action, conduct or means specifically determined by the international obligation itself which is not true of the obligation of result (Ago 1997, pp 4 et seq.). By obligation of result, on the other hand, the Commission means an international obligation which merely requires the State to ensure a particular situation – specified result – and leaves it free to do so by means of its own choice (Ago 1997, pp 20 et seq.). In other words, the distinction relates to the question whether the international obligation concerns the performance (or omission) of a particular act or the establishment/maintenance of a particular situation. The latter may also concern the avoidance of a particular situation.

---

[5] As to the development of the precautionary principle see, amongst others, Hickey and Walker (1995).

[6] Articles 20 and 21 Draft Articles on State Responsibility (ILM (1998) 37: 440)

The view of the International Law Commission that international obligations either define the conduct or the result seems to be a too simplified description on how and to what extent international law restrains the activities of States. The definition given by the International Law Commission disregards that international obligations vary as to the discretion they leave to States Parties in the interpretation of such obligations, although there exist international obligations which are defined by fixed standards.[7] The most evident examples of international obligations containing fixed standards are rules prescribing numeral reduction rates or maximum or minimum levels of emissions or immissions. Rules concerning emissions may concern the entire State – as one large pollution source – or each individual pollution source or installation. In some cases, activities are altogether prohibited; for example, according to the Convention on the Prevention of Marine Pollution by Dumping of Wastes and other Matters, 1972 (London Dumping Convention, 1972), the dumping of waste enshrined in the so-called black list is prohibited, in others, States Parties are merely obliged to cut down emissions by certain percentages or otherwise defined as for example under the obligation under article 3 and Annex B of the Kyoto Protocol. Another category of fixed standards are the ones which prescribe discharge standards, technical or product standards. Different from the percentage reductions the addressees of such standards are sources of pollution rather than a State. Discharge standards set minimum limits for the discharge of particular substances, in relation to the total emission, the time period of emission or the production capacity. International source standards have been set for pollution from ships as well as for pollution from land-based sources. Technical standards are prescriptions for technologies or operations, production, storage, or transport equipment and facilities.[8] Finally, production standards refer to the composition of products and goods either in maximum or minimum limits. Several international environmental regimes only provide a framework in which issues are to be considered and weighed against each other in order to subsequently determine what is required to do.[9] The characteristic feature of the respective obligations is that in implementing them States Parties have the discretion to consider various arguments and different external factors and somehow balance them against the obligation to implement a particular environmental standard. International norms

---

[7] ILM (1972) 11: 1358

[8] WCED Experts Group, p 55, Ebbesson, (1996, p 151)

[9] For example, under article 2, paragraph 3 lit. (a)(i) of the Protocol to the 1979 Convention on Long-Range Transboundary Air Pollution concerning the Control of Emissions of Nitrogen Oxides or their Transboundary Fluxes, 1988, ILM (1989) 28: 212. States Parties are under an obligation "... to apply appropriate national or international emission standards to new statutory sources based upon the best available technologies which are economically feasible taking into consideration Annex II ..." The words "appropriate" and "best available technologies" allow States Parties to use their discretion as to how to implement the obligation entered into. In addition, the words "economically feasible" make it possible to weigh economic considerations against environmental ones. Most international environmental treaties while leaving the States Parties some discretion on how to nationally implement the respective obligations provide some guidance on implementation or restrict the discretionary power of States. An example to that extent is provided for in article 2 of the Kyoto Protocol.

falling under this category may provide varying degrees of guidance on how to establish such balance. In spite of the discretionary power opened to States Parties as far as their implementation is concerned, the respective obligations are legally binding upon them. Apart from that, some international obligations are merely goal oriented and cannot be classified as obligations of means nor as ones of result.[10] The goal may be defined in general terms or by a quality standard to be achieved. Further, such goals may deal with emissions as well as immissions. An example for a quality related immission goal is to be found in the 1978 Great Lakes Quality Agreement between the USA and Canada. Additionally, some international treaties allow States Parties to temporarily or in part suspend the implementation of obligations if they consider this would obstruct the achievement of national policies considered to be particularly important.[11, 12] In this spirit, article 5 of the Montreal Protocol takes account of the situation of developing countries whose annual calculated level of controlled substances is limited. Those States are entitled, so as to meet their domestic basic needs, to delay their compliance with the control measure under article 4, paragraph 1 to 4 of the Protocol for ten years.

International environmental obligations defined by standards leaving the States Parties a margin of discretion how to define them may do so for different reasons. Not always is it up to each single State Party to specify the obligation; occasionally, this task is left to the States Parties as a group or to particular institutions or expert groups. In these cases the discretionary power opened is meant to provide for the flexible adaptation of international environment obligations to new developments, technologies or standards. This is a matter of rendering the respective regimes more effective rather than allowing States Parties to protect particular national interests. Accordingly, these two groups of rules should be clearly separated.

In these cases the discretion opened may be exercised only collectively, namely by the Meeting of States Parties, rather than by each State Party individually. Article 11, paragraph 4 of the Montreal Protocol, 1987, for example, provides that the Meeting of States Parties shall decide on adjustments or reductions of production or consumption of controlled substances or on enlarging the list of con-

---

[10] Leg/Ser.B/12 p. 261; this agreement prescribes "Specific Obligations" in terms of the minimum level of water quality for different substances and compounds. Environmental quality objectives have also been adopted for some substances concerning the North-East Atlantic. There are also examples in international environmental law of recipient objectives defined not in numerical terms, but as physical effects or factual situations. For example, according to article 4 of the 1974 Baltic Sea Convention States Parties should endeavour to attain several objectives by treating sewage so that the amount of organic matter does not cause harmful changes in the oxygen content of the Baltic Sea Area.

[11] The Framework Convention on Climate Change 1992. ILM (1992) 31: 818, reflects this approach by imposing different obligations upon the different categories of States. For example, article 4, paragraph 6 of this Convention provides that in the implementation of commitments a certain degree of flexibility shall be allowed to States Parties undergoing the process of transition to a market economy.

[12] Ebbesson (1996, p 85) suggests a similar, however not totally identical classification.

trolled substances (Annex A).[13] In fact, the Meeting of States Parties in respect of the reduction of controlled substances exercises legislative powers.

The process of developing a body of international environmental law, although accelerated by the United Nations Conference on Environment and Development, 1992, has, so far, not been completed. This is particularly true since one element necessary for the successful functioning of such body of law, namely effective mechanisms to ensure compliance of States with environmental obligations and the enforcement of such obligations, has not yet been elaborated adequately. The respective means developed so far are not commensurate to complexity of obligations States Parties have accepted in respect of the protection of the environment.[14]

Even the Agenda 21[15] does not contain innovations to that extent. Its Chapter 39 is limited to "implementation" and "implementation mechanisms" while the term "compliance" was deleted from the draft of what now forms Chapter 39.8. According to this Chapter, States Parties of international environmental conventions are requested "...[to] consider procedures and mechanisms to promote and review their effective, full and prompt implementation", it recommends to this end the establishment of effective reporting systems by making use of existing international institutions such as UNEP. Under the heading of "Disputes in the field of sustainable development" Chapter 39.9 encourages States to develop dispute avoidance strategies, in particular in the form of "mechanisms and procedures for the exchange of data and information, notification and consultation". No specific models were suggested to that extent though (Beyerlin and Marauhn 1997, p 91).[16] There has been a discussion to assign monitoring and possibly enforcement functions to the Commission on Sustainable Development. Under Chapter 38.13 of Agenda 21, the Commission is called upon to monitor progress in the implementation of environmental conventions. This does not, though, entail a mandate for supervising the implementation of a specific environmental regime. Nevertheless, deliberations on environmental compliance control and enforcement have recently been taken up again in international fora, amongst others in the Commission on Sustainable Development (Rest 1996, p 162).[17]

On the regional level not much more progress has been achieved regarding the development of policies dealing with non-compliance with environmental commitments. The 1993 Lucerne ECE Ministerial Declaration merely urges contracting parties to environmental conventions to adopt so-called non-compliance procedures which aim at avoiding complexity, are non-confrontational and transparent, leave the competence for making decisions to the determination of the contracting parties, allow contracting parties to consider what technical and financial assistance may be required within the context of the specific agreement, and include a transparent and revealing reporting system and procedures as agreed to by the parties (Handl 1994, p 327).

---

[13] The Second as well as the following Meetings of States Parties adopted such Adjustments, thus tightening the obligations under the Montreal Protocol. ILM (1987) 26: 1550
[14] Bothe (1996, p 13) speaks of an enforcement deficit.
[15] UN Doc. A/CONF.151/4
[16] For a review of various proposals made at different occasions to enhance the control and enforcement mechanisms see for example Plant (1991).
[17] See UN Doc. E/CN. 17/1996//L13, paragraph 8.

As far as international treaty law and resolutions and the content of resolutions of international organisations and conferences are concerned, they do not provide definite guidance as to the notions of compliance and compliance control. For the purposes of this presentation compliance means that commitments entered into by a State are fully effectuated in practice.[18] This requires actions to be taken on the national as well as on the international level. According to article 26 of the Vienna Convention on the Law of Treaties it is the obligation of each individual State to ensure its effective implementation. This means that States are under an obligation to take legislative and administrative measures necessary under their national legal system to provide for the application of such a treaty. This obligation has been specifically referred to in several international agreements,[19] in particular, in the context of establishing State responsibility.[20] Assessing whether a State has implemented treaty obligations entered into accordingly requires only the assessment of existing laws and regulations of that State. Assessing compliance, however, goes beyond this; not only that the State in question must have enacted the required laws and regulations but it must also have provided the necessary administrative procedures[21] for the enforcement of the respective rules on the national level.[22] To ensure compliance, it is necessary to establish whether the factual situation or the State actions or policies are commensurate with the international obligation.

Enforcement finally is to be understood as all the actions undertaken by States or other entities to induce or compel States to achieve compliance with environmental obligations entered into. Enforcement is the reaction to an identified non-compliance.

---

[18] The Third International Conference on Enforcement (Geradu and Wasserman 1994) provided the following definition for the notion of compliance: "Compliance is the full implementation of environmental requirements. Compliance occurs when requirements are met and desired changes are achieved, e.g., processes of raw material are changed, work practices are changed so that, for example, hazardous waste is disposed of at approved sites, tests are performed on new products or chemicals before they are marketed, etc. The design of requirements affects the success of an environmental management program. If requirements are well-designed, then compliance will achieve the desired environmental results. If the requirements are poorly designed, then achieving compliance and/or the desired results will likely be difficult." See also Sands (1996, p 49).

[19] For example the United Nations Convention on the Law of the Sea, 1982, refers to the obligation as to implementation of particular rules and procedures in several of it's provisions (for example, articles 66, paragraph 5; 69, paragraph 5; 70, paragraph 4; 211; 213; 214; 217; 222). Article 2 of the Kyoto Protocol to the United Nations Framework Convention on Climate Change, Doc. FCCC/CP/1997/7/Add. 1, gives specific guidelines in the implementation of measures and policies concerning the reduction of anthroprogenic emissions of greenhouse gases.

[20] Annex III, article 4, paragraph 3 UNCLOS.

[21] To the extent, compliance depends upon the conduct of individuals or private entities their effective supervision is a precondition for meeting international obligations.

[22] Some international agreements contain specific rules as to which actions on the national level are to be taken by States Parties to ensure compliance. For example, under the International Whaling Convention (article IX, paras. 1 and 2) Parties are obliged to prosecute and punish violators. Other international agreements are less specific in this respect.

International procedures dealing with non-compliance will have to provide ideally for the following: preventing non-compliance by co-operation, the possibility of compliance assessment, assistance in cases of non-compliance, settling of disputes, and enforcement (Lang 1995, p 685). It has to be taken into consideration that compliance is not static but a process and that the attitude towards compliance may change over time (Brown Weiss 1997, p 297).

The efficiency of a regime on dealing with non-compliance primarily depends upon the possibility to effectively monitor the respective activities of States Parties (Lang 1996, Jacobson and Brown Weiss, 1995, p 120, Lang 1998, Sands 1995b, p 141). Several mechanisms have been established to that extent but monitoring relies mostly on the assessment of reports submitted by States Parties.

As far as measures are concerned which respond to non-compliance, a distinction is to be made between instruments of command and control on the one hand and incentive-based instruments on the other hand (Brown Weiss 1997, pp 298 et seq.). This distinction has significant consequences since incentive-based instruments are often self-implementing, which means no specific enforcement action is required which States are often reluctant to take.

Until recently, the enforcement of international environmental obligations focused on mechanisms belonging to the command and control category. The primary means for the enforcement of international law are in particular unilateral and repressive instruments such as countermeasures, the invocation of state responsibility, and the various means of dispute settlement (Beyerlin and Marauhn 1997, p 73, Dahm et al. 1989, Zoller 1984).

The incentive based instruments differ widely as far as their format is concerned. Incentives may be project related or of a more general nature in which case they resemble the instrument for development assistance. Finally, new procedures have been developed recently providing for a non-controversial approach towards non-compliance.

The proliferation of non-controversial procedures responding to non-compliance has two reasons. They are meant to more appropriately reflect the different reasons for States Parties not to comply with their commitments. The attitude of a State having entered into an international environmental commitment for a particular reason – utilitarian reasons, outside pressure, in response to incentives etc. (Brown Weiss 1997, p 298) – may change and give way to different considerations leading to non-compliance. The reasons leading to non-compliance may be a matter of preference, economic or technological incapacity or inadvertence (Mitchell 1996, p 11, Bilder 1981, Bothe 1996, p 14). In particular, a State Party may opt for non-complying with a commitment if it considers that the benefits for complying do not outweigh the costs or that the benefits of non-complying exceed the ones for complying.[23] In such a situation the measures

---

[23] For the same reasons a State may decide not to adhere to a particular environmental treaty; see Blackhurst and Subramanian (1992, p 256). According to Jacobson and Brown Weiss (1995, pp 124 et seq.) the following factors are identified which may affect implementation and compliance: characteristics of the accord, characteristics of the activity the accord deals with, characteristics of the country and factors of the international environment. Some of these factors apply as much to adherence as to non-adherence to an international treaty.

taken in response of non-compliance must attempt to shift the balance in favor of compliance. If non-compliance is the result of economic or technical incapacity or due to inadvertence, measures taken in response to non-compliance might be more successful if they aim at improving the capacity of the State in question.

Additionally the proliferation of new procedures on non-compliance is brought about by the increasing complexity of international obligations (Bothe 1996, p 15). Those, in fact, vary considerably as far as their content is concerned. Although international environmental agreements become increasingly specific as to how the obligations entered into are to be implemented, they leave States Parties discretionary powers with regard to the ways and means the required result is to be achieved. The finding that a State Party is in non-compliance is, accordingly, difficult to reach and makes the respective obligation less suitable for enforcement through controversial means, such as, for example, dispute settlement procedures.

Conclusions

1. International environmental law transfers certain competencies from the national to the international level. This is due to the fact that environmental issues which touch upon the sustainable living conditions of humankind on earth are international by their very nature. Thus, international environmental law will necessarily limit state sovereignty to the extent necessary to protect the essential components of the environment.
2. The international process concerning the further development of international environmental law will continue. To do so it needs the promotion from all quarters of society. This requires a growing understanding of the earth system not alone among natural scientists but, in particular, by the society. Apart from that, efforts have to be strengthened that society not only understands but acquiesces in the consequences of environmental protection. The Convention against Desertification provides a good example for increased public participation. Any paternalistic approach – often displayed – is counterproductive.
3. Natural and social sciences will have to increase their efforts for a better mutual understanding of the respective views concerning environmental issues. Only if working conditions can be improved and established barriers be overcome, they will be able to effectively combine their efforts.

References

Ago R (1977) Sixth report on State responsibility. ILC Yearbook 1977, vol II, pp 4 et seq.
Beyerlin U (1996) The Concept of Sustainable Development. In: Wolfrum R (ed) Enforcing Environmental Standards: Economic Mechanisms as Viable Means? pp 95 et seq.
Beyerlin U, Marauhn T (1997) Law Making and Law-Enforcement in International Environmental Law after the 1992 Rio Conference, pp 73 et seq.
Bilder R (1981) Managing the Risks of International Agreement, p 112 et seq.
Birnie PW, Boyle AE (1992) International Law and the Environment, pp 2 et seq.

Blackhurst R, Subramanian A (1992) Promoting Multilateral Cooperation on the Environment. In: Anderson K, Blackhurst R (eds) The Greening of World Trade Issues, pp 246 et seq.

Bothe M (1996) The Evaluation of Enforcement Measures in International Environmental Law. In: Wolfrum R (ed) Enforcing Environmental Standards: Economic Mechanisms as Viable Means? pp 13 et seq.

Brown Weiss E (1990) AGORA: Our Rights and Obligations to Future Generations for the Environment. AJIL 84: 198 et seq.

Brown Weiss E (1992/93a) International Environmental Law: Contemporary Issues and the Emergence of a New World Order. Georgetown Law Journal 81: 675 et seq.

Brown Weiss E (1992/93b) In Fairness to Future Generations. The American Univ Journal of Internat Law and Policy 8: 19 et seq.

Brown Weiss E (1997) Strengthening National Compliance with International Environmental Agreements. Environmental Policy and Law 4: 297 et seq.

Dahm G, Delbrück J, Wolfrum R (1989) Völkerrecht, 2. edn, vol I/1, p 88 et seq.

Ebbesson J (1996) Compatibility of International and National Environmental Law, p 85 et seq.

Geradu J, Wasserman C (eds) (1994) The Third International Conference on Enforcement, pp 15 et seq.

Handl G (1994) Controlling Implementation of and Compliance with International Environmental Commitments: The Rocky Road from Rio. Colorado Journal of International Environmental Law and Policy 5: 305 et seq.

Handl G (1995) Sustainable Development: General Rules versus Specific Obligations. In: Lang W (ed) Sustainable Development and International Law, pp 35 et seq.

Hickey JE, Walker VR (1995) Refining the Precautionary Principle in International Environmental Law. Virginia Environmental Law Journal 14: 423 et seq.

Jacobson HK, Brown Weiss E (1995) Strengthening Compliance with International Environmental Accords: Preliminary Observations from a Collaborative Project. Global Governance, p 119 et seq.

Keohane RO, Haas PM, Levy MA (1994) The Effectiveness of International Environmental Institutions. In: Keohane RO, Haas PM, Levy MA (eds) Institutions for the Earth, pp 3 et seq.

Kiss A (1983) The International Protection of the Environment. In: MacDonald RStJ, Johnston DM (eds) The Structure and Process of International Law, p 1070

Kiss A, Shelton D (1991) International Environmental Law. p 10 et seq.

Lang W (ed) (1995) Sustainable Development and International Law, pp 685 et seq.

Lang W (1996) Compliance Control in International Environmental Law: Institutional Necessities. ZaöRV 56: 685 et seq.

Lang W (1998) Peer Review of Environmental Performances in International Organizations. In: Liber Amicorum Professor Seidl-Hohenveldern – in honour of his 80th birthday, p 381

Malanczuk P (1995) Sustainable Development: Some Critical Thoughts in the Light of the Rio Conference. In: Ginther K, Denters E, de Wart PJIM (eds) Sustainable Development and Good Governance, pp 23 et seq.

Mitchell RB (1996) Compliance Theory: an Overview. In: Cameron J, Werksman J, Roderick P (eds) Improving Compliance with International Environmental Law, pp 1 et seq.

Plant G (1991) Institutional and Legal Responses to Global Warming. In: Churchill R, Freestone D (eds) International Law and Global Change, p 177

Rest A (1996) Die rechtliche Umsetzung der Rio-Vorgaben in der Staatenpraxis. AVR 34: 146 et seq.

Sands P (1995a) International Law in the Field of Sustainable Development: Emerging Legal Principles. In: Lang W (ed) Sustainable Development and International Law, pp 53 et seq.

Sands P (1995b) Principles of International Environmental Law vol I, pp 17 et seq.

Sands P (1996) Compliance with International Environmental Obligations: Existing International Legal Arrangements. In: Cameron J, Werksman J, Roderick P (eds) Improving Compliance with International Environmental Law, pp 48 et seq.

Sir Palmer G (1992) New Ways to Make Environmental Law. AJIL 86: 259

Werksman J (1995) Consolidating Governance of the Global Commons: Insights from the Global Environmental Facility. Yearbook of International Environmental Law 6: 27 et seq.
Zoller E (1984) Peacetime Unilateral Remedies: An Analysis of Countermeasures

CHAPTER 5

Man's Place in Nature – Past and Future 5

HUBERT S. MARKL*

Through the evolution of the conscious mind in the human species, nature became aware of itself and can thus, for the first time in more than three billion years of natural evolution, influence and even to some degree take control of its own future development according to intentional goals. Since these goals are at the same time inevitably our own wishful visions, our species becomes not only nature's managing agent but also morally responsible for the future of nature including our own future.

This essay tries to draw conclusions from evolutionary, ecological, cultural, anthropological and moral perspectives. It not only asks about the place of human beings in nature as some kind of alien intruder, but seeks to understand human cultural evolution as part of nature as the consequent continuation of natural evolution having become not only self-organised but self-guided and responsibly self-controlled.

Humankind and Nature – Humankind in Nature

The evolutionary perspective – combining theoretical and empirical tests –, which can be the only view of nature by a present-day natural scientist, closes a very peculiar circle between nature and our concepts of nature – which thus turn out to be nature's concepts of itself. It is this circular interaction between what nature is and what nature – through the human mind – thinks of itself, which will be in the centre of my discussion on the place of the human species in nature.

Of course, I am only too painfully aware, that any discussion of the future of nature can only be even more speculative than whatever we may have to say about the past of nature. Some will say, that we can actually know almost nothing about the future of nature, because of the properties of complex, nonlinear, dynamic systems, of which nature is, of course, a prime example – the "mother of complex systems", as the Arab might say. But, we may never know, how much or how little we know, unless we try to find out. Even though we must readily concede that we will never be able to predict the future of nature reliably, it is just as true that we can foresee many possible developments well enough to make some not too improbably "educated guesses" about it.

* e-mail: praesident@mpg-gv.mpg.de

But what does the notion "Nature" mean in such a sentence when spoken by a natural scientist? For an evolutionary biologist who is not shying away from the consequences of our evolutionary insights, nature can only mean: the universe, and everything in it which is accessible to the inquisitive methods of science. This would evidently even comprise many universes – if they were to be more than theoretical cosmologists' pipedreams.

It is deep conviction – whether spelled out or not – that there is only *one and the same* nature – including the human species as far as it can be studied by science – that can be investigated by all scientists – of all ages, past and future, of all places, of all races, of whatever gender – which is the foundation of our conviction, which is just as deep, that there can be only *one* coherent body of knowledge about this nature, without any remaining contradictions at the end of our pursuit of this knowledge. In other words: we believe that there can be only *one* scientific truth with regard to this nature at least for us human beings.

At least as far as the unity of the universe is concerned, most scientists work on the assumption that there is only one ultimate explanation for all of nature. Of course, there are different perspectives, different ways of access to this one world, different subjective experiences – e.g. the artistic, the moral, the religious ones – but there can be only *one and the same scientific truth* about nature, that is at least what scientists hope and work for. Existing or seeming contradictions within a scientific view of the universe are from such a perspective remaining problems of unification to be solved by more and deeper scientific research, rather than signs of a multiplicity of incompatible scientific explanations of nature.

If we trust in such unity and continuity within nature – which is, of course, more a belief or conviction, or a postulate of practical reason, than an absolutely proven fact –, it must also follow that in reflecting and researching on the most singularly challenging offspring of nature, namely the human species, whatever is learned in reliable knowledge about ourselves – whether with the methods of the natural sciences or with those of the social sciences or the study of human culture – can only be different aspects of one and the same truth about ourselves, about what we are, where we came from and where we may be going.

In such wide perspective, the coherence or consilience of knowledge, on which Edward O. Wilson so articulately insists, is only a logical consequence of the unity of nature, of the unity of humankind and of the unity of possible knowledge of both of them. Although, as far as I see, and here I beg to differ from Wilson, this unification may not be achieved by extending concepts and insights of evolutionary biology to encompass all that can ever be found and studied in our species – just because it is without doubt a biologically evolved species – but by trying to integrate what the independent biological, social, cultural, psychological or philosophical pursuits of knowledge about ourselves have been revealing and will reveal to be true. Thus, I can only see the call for coherence as a call for uniting forces in order to overcome the divisive and misleading dichotomies which have for so long haunted our understanding of nature and of ourselves in it, for instance, as when we oppose the Animal Kingdom versus Humankind in evolution, Nature versus Nurture in the development of behaviour, Nature versus Culture in history, natural propensities versus moral rules in ethics and so on.

Only by overcoming such split-brained views of the human place in the universe, not as an inexplicably natural/supernatural twin-being, but as one being, natural down to its boots and up to its mind, will we be able to not only get a unified, a truly coherent perspective of all of nature including ourselves. This may make it necessary to look not so much in a new way on us but in a new way on nature.

It is the most important precondition of a rational, and, as such, at least to some degree predictive view on nature's future that we see ourselves not as some kind of fallen angel, alien intruder, some aberrant of deranged scourge of nature, but as its constituent and heir. And, as such, again not only as one constituent part among many others, just any arbitrarily chosen biological species, but as a unique, a quite extraordinary kind of natural species, through which nature entered an entirely new stage of its many billion years evolution; which not only participates in its future evolution like any other species, but which more and more commands and determines this future, for better or worse. In having evolved the human species, nature, as it were, has begun to take control of its own future, has become able to give it purposeful direction, has in a way become responsible for its own future development. But, mind you: all this only if we truly perceive the human species not as opposite to nature but as its most recent integral part, its own culminating invention. If the word invention may sound too intentional for some of you, I must insist that from such a comprehensive evolutionary perspective, human technological and economic inventiveness is nothing but nature's way of intentionally acting upon itself and forming its own future – maybe leading to progressive success or disastrous failure, but in both cases also bearing at least partial responsibility for such an outcome.

If that seems a strange way to argue about nature, I can only remind you that the evolutionary, the truly Darwinian perspective of putting humankind not against nature but fully inside nature has consequences of feedback circularity, which disturb our traditional ways of looking upon nature as well as those of looking upon ourselves, consequences which have hardly been thought through to their end up to now. I will try to make at least a few tentative steps in this direction. They will not be steps leading backward to a reductionist *regressus ad infinitum* but hopefully rather steps forward to a naturalist *progressus ad infinitum.*

Let me now, from such a philosophical vantage point, try to look at the future of nature (including our own species!) in five quick steps, only briefly sketching out some conceivable lines, but never forgetting that the horizons for different, unforeseen developments are wide open, that evolutionary creativity, unpredictable enough as it is, has been multiplied a thousandfold in having evolved the creative mind of the human species. It has thus literally not only entered into a new stage of creative freedom but really has created such freedom of future development of man and of nature, or to be more precise: of man (and of course: of women) in nature.

I will look at these questions from an evolutionary, an ecological, a cultural, an anthropological and finally from a moral perspective, going in rapid succession all the way from the planets to human nature. While doing this, I will, of course, restrict myself to the short-term future, because the long-term is dark. In fact, astronomers tell us, that on the long run our planet will be engulfed by the expanding red-giant sun and evaporate together with all of mankind, proving John Maynard Keynes right, who remarked that on the long run we will be all dead – except maybe those of our descendants who have escaped into outer space! But until all this happens, as astrophysicists also assure us, there are still millions of years to go for life on earth, and thus there is ample time and space for caring about the short- and medium-term aspects of nature's and humanity's future.

The evolutionary perspective

Let us first cast a quick glance at the consequences of ongoing biological evolution for the future of nature. Being relentlessly driven along at the rapid pace of human cultural change, we are always in danger of suffering from a slow-motion illusion when looking into our biological environment, which humans have for so long regarded as completely stable, as was so aptly expressed by Carolus Linnaeus hardly more than 200 years ago: *Species sunt tot, quot creavit ab initio infinitum ens.* Well, we know better now in three respects. We know first that billions of species of microbes, animals and plants have evolved over maybe more than 3 billion years, and that probably more than 99 % of them have become extinct by whatever natural causes. We know second, that due to the expansion of the human species over several thousand years both in numbers and in per capita consumption of natural resources, we are in the middle (not as some see it, only at the beginning) of one of the major extinction events in the evolution of life on earth. With human occupation and exploitation of, overall, between 10 and 90 % of the space of natural ecosystems and of a rising fraction of net biomass production all over the planet, there can be no doubt that we are taking part in and in fact causing one of the largest change-overs of biodiversity in a shorter period on the geological time-scale than life has ever experienced or suffered in the past. This will, unfortunately, be true despite everything we do to protect and maintain what is left of the biomes of our biosphere. Thus, even though we may be able to let a number of beautiful species, dear to our hearts, survive in some kind of semi-domesticated, nature-park-like fashion, they will only be sad remnants of the splendour of their life in their former natural environments, and will – for genetically inexorable reasons – on the long run never be the same again, but rather look-a-like genetically impoverished derivatives of former natural species. It might well be that some of them may only survive in the form of frozen germplasm in gene banks – or maybe even only in genome sequences on the internet –, to be revived if necessary to demonstrate what generations ago had been Siberian tigers, African rhinos, giant pandas or river dolphins. But what will definiteley have gone forever are the natural biocoenoces and ecosystems to which they belonged and which cannot survive without symbiotic member species combinations. Thus, conservation of select token species without conserving of their natural habitats and of living, sustainable species-communities, does not mean that we maintain viable parts of nature. We would only conserve odd bits and pieces from some kind of a living museum of some of nature's past.

However, we know thirdly, that at the same time that the biosphere is completely restructured under the impact of the human species, an entirely new table is set for the evolution of new species or for the expansion of existing ones which can make use of the new superabundant opportunities of 6, 8 or even 10 billion human individuals and the multibillion tons of biomass of our slave species of domesticated agricultural plants and animals. Probably never before has such a wealth of food supplies been provided so rapidly for an unlimited number of parasites and pests (as we, from our egotistic viewpoint, regard them), room for an entirely new surge of bioevolutionary creative developments, certainly not to our pleasure, but clearly bearing witness for the creative powers of natural selection. And we know very well that the harder we fight these parasites and commensals the more they will thrive and resist our means of destruction, unless they are effective enough to completely wipe out a parasite species – but even then only clearing a niche for the next inventive occupant to take possession of it.

Thus, in summary, this quick overview of clearly foreseeable evolutionary aspects of the future of nature (including ourselves) demonstrates very clearly that biological evolution will not stand still. Exactly because the impact of our super-dominating species is so cruelly effective against a large part of the existing biodiversity, it is at the same time a most effectively accelerating agent of new evolutionary development. Only for someone who would value biological nature only according to species counts – 100 species of birds or butterflies gone but 100 species of nematodes, fungus or mites replacing them – could this process of decline and rise be looked upon with equanimity. For whoever cherishes biological nature in its richness of beauty and creative complexity it is a process of heart-rending destruction and emotional loss. But since nobody knows how to return humankind fast enough by humane means from its present overblown and even still increasing state to one which could peacefully live alongside of the existing richness of living nature, and since even as massive a moral force as the Catholic Church is less than helpful in curbing the continuing growth of the human population, we will have to face the reality of an evolutionary sea-change caused by our imperialist, colonialist species which will impose on future human generations the need to fight and to come to grips with an onslaught of evolutionary forces trying to thrive on what we regard as our resources, and above all trying to thrive on that most easily accessible resource of biomass on earth, the members of the human species.

The ecological perspective

This brings me directly to the second, the ecological perspective of the future of nature. Human ecological relationships have evolved over millions of years, if looking upon the hominid family – or at least over several hundred thousands of years if looking only at *Homo sapiens* – under conditions which, for 99,9 % of the human population, no longer exist. Since the invention of plant agriculture and animal husbandry about 10,000 years ago, the human species has actively, purposefully, persistently, successfully and irreversibly remodelled its relationship to the biotic and abiotic environment, bringing about the evolutionary changes which I have just recounted. In the most advanced agricultural civilisations, artificially managed agro-ecosystems dominate over two thirds and even up to more than 90 % of the inhabitable land. These areas are more and more under the guidance of scientific knowledge and, utilising the most sophisticated modern technologies, artificially kept in a state of very high productivity of consumable biomass, characteristic for very early stages of natural ecological successions. In other words, billions of humans can only survive by continuously and artificially managing a sizeable fraction of the biosphere in a way that provides our species with needed resources, but which, by its very nature, can never become stabilised in ecological equilibrium because we cannot allow that to happen, if we want to keep them at the highest productivity.

If we now include the human exploitation of forest, river, lake or marine ecosystems and add to that the pressing need to keep at least a marginal fraction of natural ecosystems unexploited in their pristine state by actively sheltering and guarding them against further exploitative human intrusion, it must be evident for everyone considering these bare facts about the present state of the biosphere that not only those vast areas of land exploited by us directly for agricultural production, but actually almost the whole biosphere is an essential ecological resource for human survival already and will increasingly need to be purposefully and responsibly managed by our own species. It is not because we flatter ourselves in a preposterous hubris to be able to manage the whole planet better than nature, left alone, could do, but because – with all our limited insights and capabilities – we cannot escape this responsibility. A responsibility for taking care of and, where possible, cleaning up the mess which free-running human population growth and relentlessly progressing human cultural evolution – as a process of ever-increasing consumption of resources – has made of the biosphere. It is thus not a *delusion de grandeur* which forces us to accept the role of managing our way of behaving (or rather misbehaving) in the biosphere, but the sad necessity of someone who carelessly has set fire to his house and who should at least do everything to extinguish it, even if he may be well aware that he himself will never be able to rebuild the house as it had been before. "To manage the biosphere" thus does not mean an entitlement to carry on recklessly doing what comes to mind and exploiting whatever may seem exploitable, but rather to get first as clear an understanding as possible of the conditions needed for a surviving, sustainable biosphere and of what should be done and what may not be done in order to reach that goal. And secondly to organise and control our own behaviour – morally, legally, technologically, economically – in such a fashion that we really do have a chance to attain that goal.

To sum up my second point about the future of nature from an ecological perspective: the human species, whether we like it or not, has become the super-dominating species of the global biosphere and therefore has to manage the biosphere by controlling not so much the natural environment but by controlling above all humankind itself – its reproductive and propagative behaviour, its habits of exploitation, consumption and waste disposal. Thus, from an ecological viewpoint, nature has overwhelmed itself through the evolution of our species: history in that sense is only another word for ongoing evolution. As a product of nature as well as of culture, the human species is in such perspective an artefact of its own making, its own creation, in one word: self-made Man. But nature has – through the evolution of our species for the first time in billions of years – found a way to also reflect on its state and to take measures for controlling its own future development. Thus, if humans try to learn how to manage the biosphere, it is not human hubris acting against nature, it is nature's way of continuing its evolutionary path to progressive organic complexity and flexibility. Having become conscious of itself through the human mind, nature has found the means to continue its evolution beyond the reaches of biological-genetic evolution alone. Science and technology thus can be regarded as the tools by means of which nature can proceed along the way taken when giving birth to the human species and its conscious mind. While biological evolution in its so-to-speak Darwinian state is characterised by the unintentional trial-and-error adaptation of gene pools by natural selection, the rise of the human species has given nature the possibility to become self-aware of its situation and of the causes and consequences acting within itself, and thus enabled it to proceed further in a goal-directed, purposeful way through the cultural evolution of human kind. This is not in opposition to nature or against the natural laws of creation. Quite the contrary, it fulfils the very potentials of creativity embedded in an evolving nature.

Let me add one point of warning there. Most of us have heard and read about the anthropic principle, maintaining that all basic parameters of the physical universe have been set (by chance or by benign intervention) just so that life and ultimately the human species with its conscious mind could evolve; and also about the Gaia-hypothesis, suggesting that the biosphere has for billions of years been a homoeostatic, self-regulating feedback-system, which has kept conditions on our planet just about in the range of making them suitable for sustaining life on earth. Wonderful, if not miraculous perspectives of nature, but at the same time potentially most dangerously misleading ones, if we take the evolutionary message, as explained, serious. Because, if the human species is truly a natural product of the very same natural process of evolution, which has so wonderfully stabilised itself over millions and even billions of years, why then worry at all about the future? Why should the united benign forces of the anthropic and of the gaic principles suddenly fail just because *Homo sapiens* has appeared on the face of our planet? Why should mother Gaia not take care of herself with her newest great-great-grandchild on-board as ever before? What could be "unnatural" about human activities if they have so naturally evolved? In one or the other disguise we have all come across such arguments for trustfully marching on as mankind has done over the last thousands of years so successfully, putting our fate as faithfully into the hands of self-controlled mother Gaia-Nature, as ever before into the hands of the fatherly Lord.

The misconception concealed in such seemingly logical evolutionary argumentation is the neglect of the understanding, that nature in setting free the creative, insightful human mind, which is able to intentionally construct its own future, has, as it were, left the table of the evolutionary roulette-games by turning it over. In this game, nature has gambled for its own future over billions of years with great luck. But it has entered now into a completely new phase of its evolution, namely by ushering in the game of cultural evolution, where the rules of the new game are not derived by the time-proven feedback-loops of the past, but are invented continuously anew by responsible or irresponsible insightful creativity of the human mind. For this, we alone are capable. For this, therefore, we ourselves are responsible!

The cultural perspective

However, to look at human culture only from the viewpoint of setting the goal of keeping the biosphere sustainable, which is just another word for keeping it as a suitable place for humankind to survive, means to grossly underestimate what human culture or rather human cultures in their diversities can mean for the future of nature. In the first place, we should never forget that the notion of culture is derived from the Latin verb *colere*, that is from cultivating the land for growing plants for consumption by humans and their domesticated animal slaves, servants and companions. This means that culture at its very basis is not a human activity directed against nature but one of making natural productivity usable by humans. Nature in its uncultivated state is not at all a very hospitable place for human survival. While we may be most afraid of large predators hunting humans for prey, the real natural enemies are, of course, minute parasites causing infectious diseases and, above all, poisons of all kinds which plants and animals have developed to defend themselves against herbivores and predators, including us humans. The real work of cultivation meant, therefore, not so much the intensifying of the harvest but the continuation purposefully of what biological evolution had done for billions of years, selecting according to suitability, thus producing adaptations to the new, the anthropogenic environment. At its very basis, cultural evolution thus meant nothing else but the continuation of natural evolution, but intentionally for our own purposes. Therefore, it is entirely appropriate to regard artificial, man-made agricultural ecosystems from the early beginnings in Mesopotamia, the Nile, Indus or Huangho valleys, the highlands of Meso- and South America or New Guinea to the present day as nature in cultural disguise, and to see the agricultural future on the face of our earth as part of the future of nature in the geological age of the Anthropozoicum.

Just as learning from trial and error, selection by consequences under limiting constraints, gaining experience, insights and wisdom under the pressing needs of subsistence, led our ancestors to improve cultivators and cultivation processes, any future progress of the agrobiological sciences in developing higher-quality, higher-yield, more pest-resistant, environmentally less vulnerable strains of agricultural plants will continue to produce a new kind of nature out of the stocks of existing natural resources. While our ancestors, looking for the means to influence the productive yield of genetic resources, could only select what had been provided by mutational chance or accident, re-combinational genetic technologies now enable us to not only emulate the selection part of biological evolution, but also to greatly increase the variation potential from which to select. It seems difficult to regard this step as any more "unnatural" than the traditional ways of cultural selection that have been the very basis of evolution of human culture, and thus also of human nature. Therefore, agricultural biotechnology and genetic technology are only a consequent continuation of human evolution in its interaction with living nature, which makes it neither harmless or nature-proven, nor unnatural and dangerous in itself. In each and every case of its application to responsibly warranted purposes, it has to prove its worth prior to wide-scale application. To see in this a violation of natural creation is the expression of a rather limited understanding of what creation is actually about. The other way round makes more sense, the creative powers of modern biotechnology could well be regarded as the consequent continuation of that very kind of natural, evolutionary creativity which let the human species come into existence in the first place, and thus open up the opportunities for human intellectual creativity as an extension of evolutionary genetic creativity. There seems, therefore, to be more "natural" justification in the application of scientifically guided biotechnology than in the unbiologically dualistic view of the living world as being divided between biological nature on the one side and non-biological human culture, science and technology on the other.

We will have to struggle with all of our insight and inventiveness to be able to keep our population numbering in the billions alive and to keep at the same time the biosphere surviving against the odds of a world which will continue to provide us with unforeseen challenges rather than with the entitlement to a self-made garden Eden. But it is one thing to accept that *any* application of *any* technology needs responsible, moral weighing of costs and benefits and is not in itself justified just because it can be regarded as quite natural, and quite another thing to refute the misguided argumentation that scientific-technological progress must evidently be unnatural or even counter-natural because it can only be achieved through human culture and not through natural-biological processes excluding human intellectual creativity. In fact, human intellectual inventiveness and mental freedom from purely genetic programming of behaviour does not make our peculiar cultural creativity an unnatural usurpation of forbidden powers, always to be put under the suspicion of immorality. It is this very creative freedom which makes us "the moral animal", that is, the only species being in constant need of moral guidance in order to make good, responsible use of this freedom. Responsibility is not the opposite of creative freedom, but rather its inevitable consequence, almost the price we have to pay for having freed ourselves from the short leashes of genetic determination of our behaviour.

The anthropological perspective

This brings us to the fourth perspective of the future of nature, the anthropological view, that from the standpoint of the individual human being. Since every human individual is – just as is the whole species – an outcome of natural evolution, whatever we do to our biological constitution will influence the future of nature by changing our own nature. Most of us will, of course, when pondering this situation, immediately think of the newly developed possibilities of interfering with human reproductive processes and especially of genetic manipulation of the human germ line. I will return to these aspects shortly, but it seems to me that we should again step back for a moment and consider whether human rights and human dignity have only become an issue and been endangered by the recent development of reproductive and genetic technology, whereas up to now human individuals have only lived under perfectly natural and morally controlled conditions.

I think that, to pose the question in such a way, means to already answer it in the negative. One could hardly imagine anything more cruelly depriving human individuals of their inalienable human rights of life in personal freedom and in bodily indemnity, than what highly respectable, so-called "high cultures" under the close moral guidance of just as respected, so-called "high religions" have imposed at least on some, often even large fractions, often female and even majorities of their populations: slavery, witch-hunting, religiously motivated mutilation, torture, capital punishment of innocent victims; there isn't anything in even the most complete book of sadist practices which has not been applied diligently and even under the pretence of religious duty and devotion to millions of sufferers – and is often still continued to be practised even today. Thus, human cultures of the past have never shied away from inflicting the gravest damage to human nature without regard of its dignity, as little as they have from inflicting such damage on other species. They have also never hesitated – again with little respect for human freedom and dignity – to subject growing girls and boys to the most severe pressures of indoctrination in order to make them conform to a society's habits and norms, which, from a standpoint of enlightened human dignity, can only be judged as abominable deprivations of the most basic human rights.

It would, therefore, seem appropriate, when dealing with imagined or real new dangers for human dignity as a result of the new possibilities of reproductive and genetic technologies not to be carried away with the impression that their indisputable dangers, just because they are new and hitherto unknown, are in any way more serious and despicable than the long-existing arsenals for depriving human individuals of their freedom and dignity. This is evidently not so for thousands and thousands of healthy boys and girls who owe their lives to artificial conception and embryo transfer, more often than not more dearly loved boys and girls than many born unexpectedly and unwanted under the most natural circumstances. And I can also not see that the future of human nature as one respecting individual rights and dignity is unacceptably comprised if some of the most severely genetically malformed and permanently disabled individuals are not exposed to a cruel, short life and early death by trying to avoid such a development, which to call natural – although logically correct – would mean making what occurs by the forces of nature the ultimate moral imperative, a form of naturalistic fallacy of jumping from "is" to "ought" which moral philosophers have long taught us to avoid. And even if the example of the possible cloning of human individuals comes to mind – which most of us would want to see precluded by force of law for good moral and social reasons – it should be remembered well that – in case of success – the worst that could come out would be a completely new human being with all its human rights and with the undiminished entitlement to the respect of its dignity.

It seems necessary to emphasise strongly here that such argumentation is neither aiming for any kind of eugenic amelioration of the human gene pool nor can it in any way – neither logically nor morally – support such mistaken eugenic goals. It is also not convincing that the increasing costs of healthcare budgets would force us to earnestly consider such eugenic cleansing of the human gene pool, for the following reasons, among others:

- The increase of costs of healthcare is caused far more by rising medical expenses during the last years of ageing patients than during the first years of genetically severely handicapped individuals;
- most seriously genetically handicapped human beings die in their first months or years of life – well before reaching reproductive age, even with best medical treatment, and thus cannot contribute to the genetic load of human population;
- of those, who do survive to reproduce, one person's handicap more often than not is another person's special gift;
- an overall degradation of the human gene pool has never been proven and seems highly improbable, since its genetic composition is mostly dominated by the more than 90 % of human population receiving little medical attendance;
- in the remaining 10 % of the human population, who can afford to receive such special medical attendance, costs of healthcare seem not so much driven by genetic deterioration but rather by rising numbers of medical practitioners.

It seems therefore not only immoral and illogical to turn to eugenic gene pool management of the human population, but also quite improbable that there is any real danger that the human species could first go broke and than maybe even extinct because of exaggerated healthcare efforts for genetically severely handicapped new-born individuals.

Therefore, it seems to me that, while we have good reason to look with sober judgement and full moral responsibility on all these new or sometimes not-so-new perils of scientific and technological progress, we should also not be carried away too readily by the "sweet temptation of feasibility". There is also every reason to open-mindedly consider what advances in human biology might have to offer us in order to avoid or cure suffering and to help human beings live fulfilled and non-deprived lives. Not everything coming into the reach of feasibility needs to be done, and there is certainly much that should not be pursued at all out of respect for human freedom and dignity. But it seems to me that this aspect of the future of nature, namely the future of our human nature, is so much more determined by the powerful processes of cultural evolution, especially by the almost unfettered "cloning" of misleading ideas, superstitions, prejudices, anxieties, chauvinist beliefs and unethical desires by means of indoctrination, suggestive advertisement and immoral (and sometimes even pretended moral) coercion, that we should do well to not worry so much about the most novel but rather about the most perilous and quite often quite ancient dangers to human nature in the future of nature.

The moral perspective

That takes me to my fifth and final consideration with regard to the future of nature. I have, I hope, been able to make it clear enough why I believe that the future of nature at least from now on, but actually for quite some time already, cannot be treated as if it were independent from the future of what is tradition-ally regarded as being beyond nature: human life and culture, the world of human thought and imagination, the mental world in which everyone of us lives just as self-evidently as in the natural surrounding world. In fact, none of us would for even one moment deny that it is this mental world of the human mind that is the very core of our human nature. Compared to it, everything else which can be called upon to make us distinctive from our animal relatives – be it an upright posture or hairless skin, feeding or mating habits or whatever else – seems insignificant, with the only exception of language, but then it is, of course, our capacity of creatively using language which, more than anything else, is the tool with which we can become aware of the mental worlds of others and with which we can give others access to our private theatre of imaginative plays, crea-tive narratives, and original thoughts and feelings.

Since we owe this capacity for conscious thoughts and emotions and their expression to our supersized brains which are the outcome of the process of the natural evolution of our species, one cannot focus on the future of nature without ending up focussing on what this mental world, and its further development may mean for the future of nature.

If it is true that nature, through the evolution of the human species, and particu-larly the human brain, has found a way to reflect upon its past and present condi-tions, and even to some limited degree on possible future developments; if nature in one word has become conscious through human consciousness and free to act through human freedom to act, it has brought itself to the point where the future for nature is no longer limited according to the laws and boundary conditions exis-tent in our universe, but to the point where the future becomes a potential to be realised, a goal to be achieved and worked for, not made in the sense of being completely under the control of performance, but very much so in the sense of imposing responsibility on conscious actors for their conscious actions. That means to say that "to make a future" is very different from letting it just occur by behaviour; namely by being obliged to act according to reason and by being responsible for what is made. Of all creatures in nature it is only our human species through which nature can act in such ways. Therefore, because we have evolved the capacity to not only behave but to act purposefully and to thus "make a future" for ourselves and for nature, we cannot escape the responsibility to do so.

It is this fact of our nature which makes us dependent on moral guidance, that is, on guidance how to pass judgement on what is good or bad and on how we should lead our lives. To ask about both the sense of and the sense in our lives is just another way of asking for moral guidance in order to know what is good and what is worth striving for, to work and to suffer for. Nature has not completely set us free without giving us a well-equipped package of desires, emotions, hopes and fears, longings and needs in order to keep us going and searching. But with all these inborn emotional driving forces we are like a self-starting locomotive machine fully equipped with engine and fuel but much in need of a clear vision of where to go and of a map telling us how to get there. Clearly, there are some who feel that to keep this engine running and to experience the good feelings going along with it may be all there is to make life worth living and which there-fore already answers the question about the meaning of life. But most of us want

– at least at times – to reach beyond that, to define a goal to reach for and to manage to do so during the course of one's life. Evidently, this is the point where the biologist should stop and hand over the leading role to the moral philosopher or the religious teacher, who try to give answers to the questions of moral guidance. But even then we should remain aware, that we are by doing this not leaving the boundaries of nature, but that we rather give the "moral sciences", as the humanities and social sciences used to be called so aptly, their proper place in discussing the future of nature and humanity's place in it. Because any answers which humans may give to such questions will always at the same time and by necessity have to be scientific *and* moral answers. Only both together can fully grasp humanity's place in nature, in the future just as well as in the past.

PART II

Focus: II
Water in the Earth System: Availability, Quality and Allocation in Cross-Disciplinary Perspectives

CHAPTER 6

Global Alteration of Riverine Geochemistry under Human Pressure

6

M.-H. Meybeck*

Riverine geochemistry and material fluxes have already been much altered at the global scale by agriculture, deforestation, mining, urbanisation, industrialisation, irrigation and damming which have generally appeared in this order. The continental aquatic systems (CAS) are now affected by hypoxia, eutrophication, salinisation, and contamination by nitrate, metals and persistent organic pollutants. The historical development of these impacts is now being reconstructed by sedimentary archives or assessed by direct measurements for the last 100, 50 or 30 years. The societal responses to these water quality issues can be described by half a dozen typical strategies and their time scales, which generally spread over more than 20 years are controlled by both environmental and societal inertia. Major differences in environmental control efficiencies are expected between industrialized countries, for which control measures adapted to each occurring issue have been gradually set up over last 50 to 100 years, and fast developing countries which are facing these issues in much shorter periods.

General human impacts on aquatic systems

River fluxes are studied for three main purposes: (i) as indicators of chemical weathering, soil leaching and mechanical erosion, (ii) as inputs to oceans, regional seas and internal areas, (iii) as indicators of human impacts on Earth systems.

Rivers are exposed to two major types of global changes that may affect their geochemical composition and fluxes: (i) climate change which controls riverine hydrology, chemical weathering, and soil leaching, (ii) direct anthropogenic changes as agriculture and deforestation, mining, urbanisation, industrialisation, irrigation, damming. Climate change is still a slow process and its impact on recent riverine chemistry – over the last one or two hundred years – it still limited while direct human pressures have resulted in marked qualitative and quantitative modifications of continental aquatic systems (CAS) at the local to global scale. These local effects concern water and sediment chemistry, physical habitat, hydrological regime and will have a marked impact on aquatic biota. Most of these modifications now affect the entire Earth System, of which river fluxes to the oceans is an important segment recognised since Garrels and Mackenzie

* e-mail: meybeck@biogeodis.jussieu.fr

(1971). Early attempts to differentiate pre-anthropic fluxes from the present-day situation have pointed out the influence of Man at the global scale (Meybeck 1979, 1982) who should now be considered a major driver of the riverine segment of surficial geochemical cycles.

Climate change and direct human pressures on CAS have resulted in global issues of many kinds that may hinder our socio-economic development in many fields (see table 1): (i) human health, through water-related diseases as malaria and bilharzia, and water-borne diseases as cholera, and through the development of toxic algae linked to eutrophication, (ii) water availability, through enhanced evaporation, particularly during irrigation, through water transfer, through aquifer overpumping, (iii) water quality, through diffuse and point sources of pollutants from agricultural, urban, industrial, and mining sources, (iv) global carbon balance, through the carbon storage of particulate carbon in floodplains, lakes and reservoirs, through the possible destorage of organic carbon during permafrost melt, through enhanced silicate mineral weathering by soils CO_2 at warmer and wetter climatic conditions, (v) fluvial channel morphology, through changes in sediment balance (erosion/transport/retention), (vi) biodiversity of CAS, through the general loss of pristine water courses, segmentation of river networks by dams, the habitat and water quality degradation, the river system cutoff from the coastal zone or from their internal end points (Amu Darya and Aral), (vii) coastal zone changes resulting from all previous impacts as changes in the hydrological and sedimentological balances, increase of some nutrients loads (N, P), increase of toxic metals and persistent organic pollutants (POP's).

The historical development of human impacts on CAS started in the hydraulic civilisations of Egypt, Mesopotamia, Indus and Amur Darya basins, and in China (Mainguet 1999) as a result of deforestation, agriculture, or hydraulic works. Severe neolithic and Bronze-Age mining contamination is also known as in the Rio Tinto basin (Leblanc et al. 2000) in southern Spain. This degradation was first restricted to the very local scale (basins < 1,000 km^2). In large basins (0.1 to 1 Mkm^2), impacts are not found before the Industrial revolution – with the noted exception of early hydraulic control of Chinese rivers. Figure 1 postulates this inverse relationship between pressure and environmental impacts with increasing basin size, and its evolution with increasing human development. This trend is valid for point sources of pollution like cities, industries and mines but may not apply to diffuse sources.

Original water chemistry, water quality and pollution

The original state of CAS are difficult to determine. In large basins where the water chemistry at mouth corresponds to areas exceeding 100,000 km^2, the ranges of most water quality attributes spread over one order of magnitude, and two orders of magnitude for the Total Suspended Solids (TSS) (Meybeck 1996). When considering smaller basins (area < 10,000 km^2) this spatial variability is enhanced, and most attributes range over two orders of magnitude. The most dilute river waters in Central Amazonia, for example, are 100 times less mineralised than some Central Asian rivers.

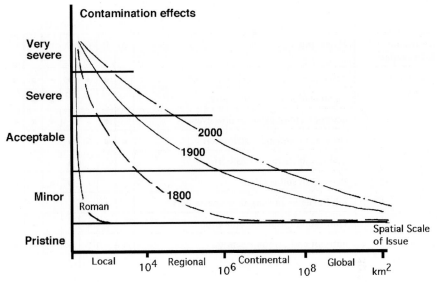

Fig. 1. Schematic evolution of contamination of continental aquatic systems at various scales from Roman time to Present.
Source: adapted from Meybeck et al. (1989).

Human activities require a certain chemical composition for water, or particulates. Such water quality requirements may vary according to the different uses (e.g. irrigation, livestock watering, drinking water, electronic, textile, brewery). They may also vary in time, and differ from one country to another (e.g. eutrophication criteria in Scandinavia and Canada vs Southern Europe). Drinking water standards established by the World Health Organisation are used, but not always enforced, globally. In many regions of the globe the natural water quality is inappropriate for many human uses, particularly in groundwaters. Such situation should not be termed pollution which is a commonly used term, yet sometimes confusing. Many river basin managers are combining several different constrains into one single concept of "water quality": the compliance with water use standards, the modifications of water chemistry from pristine state – often difficult to establish – and the compliance with general aquatic biota standards which seldom take into account the local original water chemistry. Pollution should be kept for the most severe degradation of continental aquatic systems resulting from human impacts only.

The evolution of the chemistry and quality of aquatic systems is difficult to address through direct measurements since water quality surveys are relatively recent in History, one hundred years for the longest record but in most cases only 50 to 30 years (Meybeck et al. 1989). A few case histories of water quality issues listed in table 1 are illustrated below.

Table 1. Major global threats to continental aquatic systems and related issues

Environmental state changes	Major impacts	Global Issues						
		A	B	C	D	E	F	G
1. Climate Change	development of non-perennial rivers		●	●	●	●	●	●
	segmentation of river networks					●	●	
	development of extreme flow events		●			●	●	●
	changes in wetland distribution	●	●	●	●		●	●
	changes in chemical weathering				●			●
	changes in soil erosion				●	●		●
	salt intrusion in coastal ground waters		●					
	salinisation through evaporation		●	●			●	
2. River damming and channelling	nutrient and carbon retention				●			●
	retention of particulates				●	●		●
	loss of longitudinal and lateral connectivity						●	
	creation of new wetlands	●		●	●		●	
3. Land use change	wetland filling or draining			●	●		●	
	change in sediment transport				●	●		●
	alteration of first order streams					●	●	
	nitrate and phosphate increase	●		●	●			●
	pesticide increase	●		●				●
4. Irrigation & water transfer	partial to complete decrease of river fluxes					●	●	●
	salinisation through evaporation		●	●				
5. Release of industrial and mining wastes	heavy metals increase	●		●				
	acidification of surface waters			●			●	
	salinisation	●		●			●	
6. Release of urban and domestic wastes	eutrophication	●		●	●		●	●
	development of water-borne diseases	●						
	organic pollution	●		●			●	
	persistent organic pollutants and metals	●		●				●

A: human health, B: water availability, C: water quality, D: carbon balance, E: fluvial morphology, F: aquatic biodiversity, G: coastal zone impacts. Only the major links between issues and impacts are listed here.
Source: (Meybeck, 1999).

The Thames history of oxygen balance

The dissolved oxygen concentration has been measured over the last 150 years in the Upper Thames estuary (see fig. 2). This record is probably the oldest continuous quantitative analysis of a CAS. In the 1850's the organic pollution from greater London (2.75 million people) was such that all O_2 was depleted and no fish of any kind could survive. The last salmon was caught in 1839. Between 1830 and 1871, several cholera outbreaks were responsible for more than 30,000 human casualties (Schwartz et al. 1990). A massive urban sewage collection was undertaken during the late 1800's to cope with the growing population (4.75 million people in 1880). As a result of this environmental management, the first of its kind (soon followed by other big cities as Paris), the estuarine quality was restored and the O_2 saturation went up to 50%, a satisfactory level for most fish. With the continuous increase of the urban pressure (8 million in the 1940's) the sewage treatment rate was insufficient again and a second hypoxic phase ($O_2 < 2$ mg/L) was noted from 1900 to 1960 until another massive effort of waste collection and treatment was undertaken. In the 1970's the 50% saturation point was reached again.

This exceptional case story points out two important features: (i) very severe environmental degradation have occurred in Western Europe during the 19th century at a time of very rapid industrial and demographic growths when no environmental control was present, (ii) environmental quality trends may be multicyclic: the present-day situation may not always be the worst.

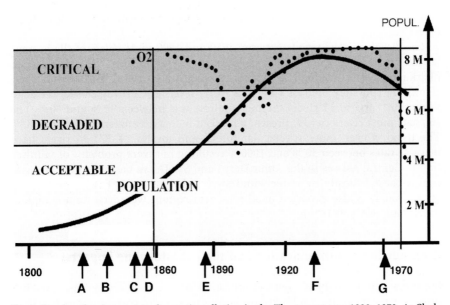

Fig. 2. London development and organic pollution in the Thames estuary 1800–1970. A: Cholera, B: Last salmon, C: "Great Stink", D: Sewer completion, E: Primary treatment (Beckton/Crossness), F: Modgen, G: Full secondary treatment (Beckton).
Source: adapted from Schwartz et al. (1990).

The global salinisation trend of rivers

The salt content of rivers in natural conditions depends on rock type and climate: basins rich in rock salt outcrops, gypsum and carbonated rocks have the highest natural water mineralisation while waters draining granite, gneiss and sandstone are the least mineralised. In addition in semi-arid climates, where evaporation exceeds precipitation, surface waters are gradually concentrated. The combination of these factors may lead to high total dissolved solids (TDS), as in Central Asia.

Most human activities result in an increase of major ions, in surface water, particularly Na^+, Cl^-, SO_4^{--}, K^+, either through direct inputs or through enhanced evaporation in reservoirs and in irrigated fields. In the Murray River basin (Australia) the headwaters have a quite low TDS content (30 mg/L). TDS gradually increase downstream and reach 3,000 mg/L at mouth in the 1960's, as a result of irrigation returns and natural saline groundwater inputs, a level by far exceeding the 800 mg/L value considered as a water quality objective for most uses (Chilton 1989). Similarly the Cl^- levels of Rio Grande headwaters in Colorado State are only 7 mg/L compared to 500 mg/L in downstream Texas, a level exceeding the 200 $mgCl^-/L$ objective.

The Cl^- trend in the Rhine is documented for the last 100 years. It has increased from an estimated pristine level 5 mg/L in 1900 to around 200 mg/L for the last 30 years, due to the brine releases from Lorraine salt mines and of Alsace potash mines: from 10,000 to 15,000 t NaCl are released daily into the river from the potash mines. After a severe conflict between the Rhine countries, the French mining companies agreed to adjust their NaCl inputs to meet the 200 $mgCl^-/L$ objective.

The dissolved salts trends between 1950 and 1990 in 41 rivers of the Former Soviet Union is very much representative of the salt issues at the global scale (Tsirkunov 1998). Calcium (Ca^{++}) is the least sensitive: in 27 rivers its trend is not significant (possibly because Ca^{++} and HCO_3^- levels are already closed to calcite saturation), and in 11 rivers the increase rate was between 40 % and 80 %. For sulphate most rivers (n=39) present a marked increase between 20 % and 500 %. The trend is even more spectacular for Cl^- and for $Na^+ + K^+$: 36 rivers have increase rates between 50 % and 1000 % resulting in severe problems of salinisation in Central Asia as in the Amur Darya and Syr Darya basins (where salinisation is caused mostly by water withdrawal and irrigation). The salinisation of groundwaters is also expected in all semi arid regions were irrigation is rapidly developing (Chilton 1989).

River nutrients, eutrophication and their coastal zone impacts

Since the 1950's and 1960's, eutrophication of lakes and reservoirs has spread in most continents as the result of phosphorous increase in surface waters (particularly from untreated collected urban sewage and from the use of phosphorus detergents). Vollenweider's (1989) global survey identified that 65 % of OECD lakes were eutrophied.

River eutrophication trends have been more recently noted in slow flowing and nutrient-rich (Rhine, Seine, Loire) or in impounded rivers (Volga). Chlorophyll maxima in these rivers commonly exceed 100 µg/L and reach 200 µg/L, levels that would qualify them as hypertrophic in lakes and reservoirs.

Several major issues are related to river eutrophication: (i) a marked daily variation of river chemistry (ΔpH up to 1.5, ΔO_2 saturation of 100 %) which can stress some fish species, (ii) the production of highly labile organic matter, (iii) the complete modifications of elemental ratio N : P : Si in waters reaching the coastal zone. In the Seine the Si : N ratio was 48 g/g in pristine conditions, it is estimated to be only 0.9 today. In the Mississippi the Si : N ratio was around 10 g/g in 1900 and now varies between 2.3 and 4.5 g/g (Turner et al. 1991). Silica may become a limiting factor in some coastal environments resulting in a shift of algal species as in the Danube, where silica depletion is attributed to the Iron Gates reservoir (Humborg et al. 1997).

Coastal zone impacts of river eutrophication and nutrients inputs are severe. In the Loire and Seine estuaries hypoxic conditions have been observed in summer over the last 20 years. Offshore from the Mississippi, the mid-summer hypoxic area has increased from 8,000 km^2 in 1985/92 to 18,000 km^2 in 1993/99 (Rabalais et al. 1999).

Both phosphate (PO_4^{-3}) and nitrate (NO_3^-) increases are observed in most rivers exposed to human pressure, a general process which has been termed the opening of nutrients cycles (Billen et al.1998). Their sources are multiple. Since the 1950's the use of nitrogen and of phosphorus, both as fertilisers, and in the food, detergent, and other industries, have resulted in rapid increase of riverine N and P fluxes now exceeding the pristine levels by a factor of ten in some Western Europe rivers. Both nutrients present complex biogeochemical cycling and transformation in CAS, including retention of particulate forms in soils, lakes, floodplains and reservoirs, biological uptake and regeneration, and escape to atmosphere (for N_2O).

The nitrate contamination illustrates the difficulty of handling diffuse pollution sources. In the Seine basin, the fertiliser origin of nitrate was revealed by natural tracing with 15 N (Mariotti et al. 1975). It took 9 years of scientific and policy discussions before both French Ministries of Environment and Agriculture provided the first advisory guidelines in the early 1980's. In the early 1990's some subsidised farmers were the first to use these to secure the recharge area of major commercial spring bottled waters. Despite numerous European and French regulations, fertiliser use has not really decreased in the Seine basin and nitrate is still increasing. The nitrate issue is not likely to be solved in the next decades given the residence time involved in soils, unsaturated zone and ground waters.

On the global scale, nutrients inputs to the oceans by rivers have already increased 2.2 times for nitrate and 4 times for ammonia according to a recent data base compiled from 200 world rivers representing half of the river discharged to oceans (Meybeck and Ragu 1997). The pristine average levels in world rivers are estimated from this data base to 0.11 mgN/L for NO_3^-, 0.023mgN/L for NH_4^+ and 0.018 mgP/L for PO_4^{-3}, values which are close to a previous estimate (0.10, 0.015 and 0.010 mg/L for N-NO_3^-, N-NH_4^+ and P-PO_4^{-3}, respectively, Meybeck 1982).

Fig. 3. Trends of nitrogen and phosphorus species in the Rhine River (data from the International Commission for the Protection of the Rhine, Koblenz, Germany).

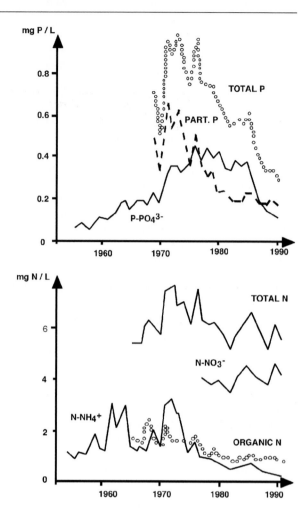

This global figure masks enormous disparities in nutrients impacts. The most polluted waters originating from Western Europe and part of North America represent less than 10 % of water fluxes but half of nitrate and ammonia fluxes.

Nitrate abatement has been not yet documented in any of highly impacted rivers but ammonia abatement is noted wherever source of the organic pollution is correctly addressed as in the Rhine (see fig. 3).

The phosphorus issue is more complex. In the Rhine basin, sewage collection and secondary treatment in the 1970's removed most of the excess particulate P (see fig. 3) but the decrease of phosphate has needed an additional control such as tertiary treatment, and a severe restriction or a complete ban (for Switzerland since 1985) of P detergents. In France, where the phosphorus industry had a more successful lobby in the 1980's and 90's, phosphates levels are comparable to those of the Rhine 20 years ago and have just been stabilised.

Metals, an old story, a variable success

Prehistoric impacts from human activity can be observed in cores taken in river deltas or flood plains. In the Rio Tinto, South Spain, Leblanc et al. (2000) have found evidence of neolithic mining dated at 2530 BC with levels of As, Cu, Hg and Pb exceeding the natural background by hundreds of times. Similar enrichments are now found in river basins affected by present-day mining activities but at much larger scales. Contamination history by mines can be complex: in the Rio Tinto other peaks are noted up to the present mining activity, between which the basin has only partially recovered. In the Meuse basin (Belgium), the contamination started in 1650 and peaked in 1900 with levels reaching 10 to 50 times the local background (Rang and Schouten 1988). In the Lot river (France), where the Riou-Mort zinc mine was in operation from 1840 to 1980, 0.5 Mt of contaminated sediments (exceeding sometimes 10 µg/g Cd) containing 200 t of cadmium are presently stored in river bed sediments. They are gradually moving downstream and contaminating the Gironde estuary and the Marennes oyster farms some 400 km away (Blanc et al. 1999).

Severe contamination is however well documented in numerous former industrial sites in the US, and in Former Soviet Union. According to Zhulidov (1998) there is clear evidence in river sediments of very severe Cu and Ni contamination from "Severonickel" in river sediments of Kola peninsula (NW Russia), of Cu and Zn contaminates from the "Ust-Kamenogorsk" and "Irtysh Polymetal" smelters in Kazakhstan, of Cu and Ni contaminates from Norilsk smelter in north west Siberia (Russia). The expansion of mining in the last 150 years has spread the potential metal contamination to very remote places as Irian Jaya, North West Territories, upper Zaire river basin.

In non-mining areas, very high metal contamination can also be found due to use in domestic, urban and industrial activities. Metal contamination can be

Fig. 4. Metal contamination in Seine River recent flood deposits (Horowitz et al. 1999) normalized to average background concentrations (BK). The differences in the ratio basically track from upstream (BK) to downstream sites (DD). Paris city is located in UR (see text). Pop. dens.: population density of subbasins upstream stations (people/km²).

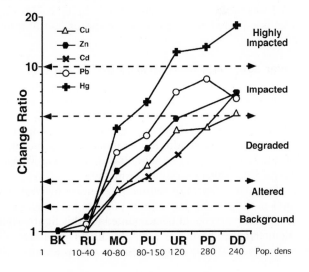

observed in the river flood particulates of the Seine basin (see fig. 4). Pristine reference levels have been measured in selected forested small catchments (see fig. 4, BK) with population density around 1 person/km². From the rural basin (RU 10 to 40 p/km²) to the mid order streams (MO 40–80 people/km²) the metal contamination slowly increases with increasing population density. A strong impact is noted downstream of the release of Paris city treated sewage (PD 280 people/km²). Industrial activities add further contamination between Paris and the coastal zone (DD 240 people/km²) while the population pressure remains constant. In the Seine basin (65,000 km², 17 million people) cadmium, zinc, lead, copper and mercury are the metals most sensitive to human activities, while nickel, vanadium, lithium and cobalt are least (Meybeck 1998, Horowitz et al. 1999).

On the global scale the impact of human activities on metal transfer to the oceans is very difficult to assess. After surveying dissolved metals in rivers for 20 years most water authorities in North America and Europe have discarded their data-sets, suspect of contamination from sampling to analysis, and have chosen the analysis of particulates.

Persistent Organic Pollutants: a growing concern, a poorly known situation

Some organic pollutants are highly resistant in the environment and can still be detected many years after their primary use. They may be xenobiotic (polychlorinated biphenyls (PCBs) or DDT), or already exist as traces in natural conditions (the polyaromatic hydrocarbons (PAH)). Many of these products are poorly soluble and are linked to particulate transport pathways from soils to river sediments.

The analysis of POP's in continental aquatic systems is much less common than of metals. Most of the reliable information is found in sediment core data. A survey of estuarine data by NOAA (Valette-Silver 1993) showed various patterns: PCB's and DDT were decreasing in Puget Sound and Hudson estuary while PAH was increasing or stable. In developing countries, severe POP's contamination is expected due to present use and/or poorly enforced regulations.

River damming: from local to global impacts

It has been estimated that about 40 % of river flow to the oceans is already intercepted by reservoirs (Dynesius and Nilsson 1994, Vörösmarty et al. 1997). Major dams have now been built on most major rivers in all continents (Columbia, Nelson, Missouri, Colorado, Rio Grande; Caroni, Tocantins, Parana; Volta, Senegal, Niger, Zambezi, Nile, Orange; Indus, Yang Tse Kiang, Huang He, Ob, Yenissei; Volga, Don, Dnieper, Danube; Murray).

The first estimate of sediment retention by world major reservoirs is at least 16 % of the present-day sediment flow to the ocean (Vörösmarty et al. 1997). If all small reservoirs could be accounted for this figure would be much higher. In the Mediterranean basin, for example, the total sediment discharge was estimated to be around 620 Mt/y in the 1940's: It is now around 180 Mt/y. Lake Nasser alone

Fig. 5. Trend in water discharge (I) and sediment discharge (II), 106 t/y after Hoover Dam construction in 1936.
Source: Meade and Parker (1985).

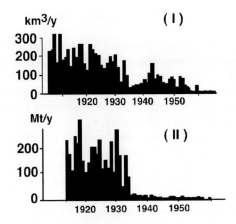

stores 120 Mt of Nile silt annually. The global reservoir construction trend masks the impact of land erosion, long term fluvial sediment transport records at river mouths, usually displaying a decrease in sediment discharge as in the Colorado River (see fig. 5). Other impacts of reservoirs at global scale include the fragmentation of riverine habitat and nutrient uptake (Dynesius and Nilsson 1994, Humborg et al. 1997) and particulate pollutant retention.

Irrigation: a XXIth century challenge

In the semi-arid and arid zones rainfed agriculture is limited; and irrigation is a common practice. Irrigation has dramatically increased in the last 50 years, and is closely related to the development of dams and reservoirs. An estimate of the total area irrigated, 2.4 Mkm^2, is given by Shiklomanov (1998). In many river basins waters are already so heavily used by irrigation that severe reduction (20 % to 50 %) of river discharge to oceans is now common (Orange, Rio Grande, Murray, Ebro). This may result first in a complete cessation of river flow during dry periods, as is observed for the Huang He. When all resources are used for irrigation, and sometimes for megacity supply (as for Los Angeles), the river effectively dries up and there is less than 10 % of annual discharge left to the ocean (see fig. 5), as for the Colorado, the Nile and for the Amu Darya. For certain regions the decrease of water input to oceans can be estimated: in the Mediterranean it dropped from about 510 km^3/y to less than 200 km^3/y. In the Former Soviet Union the flow of Kuban, Dnieper, Don, Ural, Terek, Sula, Kura and Ily has been reduced by 27 to 44 % and the Volga by 10 % (Chernogaeva et al. 1998).

At the global scale, the water discharge from continents to oceans has already been reduced by 6 % (about 2,500 km^3/y). Most of this consumption is attributed to irrigation (Shiklomanov 1998).

Societal response to water quality issues

The water quality trends observed in rivers and lakes result from human pressures, hydrosystem response to pressure, social and societal awareness, political decision, financial and technical instrumentation for environmental control, as well as hydrosystem response to environmental control. Based on previous examples few major types of river basin management strategies to control water quality issues can be described (see fig. 6).

Strategy A (**unnecessary management**) is observed either in low pressured river basins or for minor changes from natural concentrations (CN) which are irrelevant to economic uses or to aquatic ecosystem stability (e.g. K^+ increase in all rivers). In both cases there is no environmental management although regular controls should be made.

Strategy B (**precaution management**) corresponds to successful management well ahead in time which targets the recommended level (CR) so that the environmental impacts and economic losses are both kept at the minimum levels. If the control is not planned early enough some degradation may occur. In eutrophication control both results have been observed.

Strategy C (**maximum impact management**) corresponds to a choice of the maximum acceptable degradation (CL). This can be found in countries or basins with limited economic means or as the result of negotiations between various

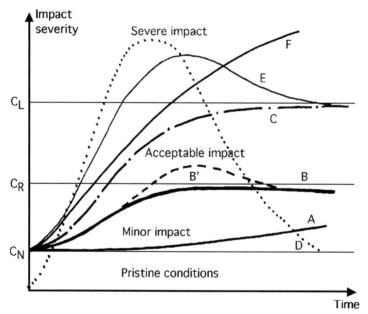

Fig. 6. Typology of river basin management strategies for water quality. C_N, C_R, C_L: natural, recommended and limit concentrations (see text).

stakeholder sharing the same resource. The Cl⁻ regulation in the Rhine is a typical example of such strategy C.

Strategy D (**total ban**) combines first a non-management of the issue which goes up to a severe impact level then followed by a major control with a zero impact target as for DDT and PCB bans in Europe and North America.

In strategy E, (**pollution regulation**) after a period of absent management and severe impact, the targeted level is the maximum acceptable degradation. Nitrate pollution regulation in Brittany rivers, one of the European regions most affected by this issue, follows this rule: less contaminated waters are now stored in reservoirs and released into rivers to dilute nitrate at the WHO drinking water limit.

The **laissez-faire** strategy F represents either the lack of environmental control, as in Minamata, or its lack of efficiency. Such a strategy is of course unsustainable. The needed environmental reclamation which may come decades after the removal of the original pressure, is realised at highest costs.

If natural concentrations (CN) already exceed the targeted concentrations for human uses the water quality management strategies may require specific treatment of the natural water resources.

A conceptual framework for understanding the stages of societal and socioeconomic responses to degradation of continental aquatic systems can be proposed, depending on pressure intensity, time constants and varying societal responses (see fig. 7):

1. **hydrosystem reaction to contamination**, depending on system size and contaminant pathways;
2. **impact detection** by water users, scientists, specific citizen groups of the change to the hydrosystem;
3. development of **societal awareness**: time for the development of a general consensus about the issue;
4. **policy lag**: time for administration or politicians to decide on action;

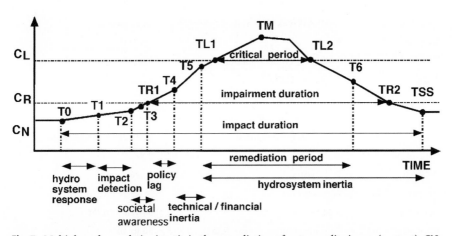

Fig. 7. Multiple and cumulative inertia in the remediation of water quality issues (see text). CN, CR and CL: natural, recommended and limit concentrations.

5. **financial and technical lags:** time to fully enforce the decisions;
6. **hydrosystem reaction to remediation** or control measures.

These stages are here illustrated for an effective water quality control as for strategies B, D, E. Depending on the timing of detection, impact duration and remediation effectiveness, various threshold levels can be reached. If the environmental control is effective enough, a limited level of degradation is reached, followed by an improvement phase. When partial control strategies (Strategies C and E) are in practice, the new environmental state stabilises near the critical level CL. If control is more effective, a critical period (TL1-TL2) can be defined during which severe ecological impacts, human health impacts, or economic restriction are observed. If the recommended level (CR) is again reached, the impairment period (TR1-TR2) characterises the total duration of impacts.

In reality the full recovery cycle is rarely found, the common management types being C, D, E and F. The history of CAS reveals many contrasting examples, where some of these evolutionary stages are very reduced or absent. The hydrosystem reaction mostly depends on water renewal. When the impact is linked to soil or sediment contamination and water bodies with long residence times (ground waters, large lakes), the hydrosystem reaction may take decades and more.

Table 2. Milestones and time scales of environmental restoration: examples of the Tokuyama Bay (Japan) and Lake Léman (Switzerland, France) (see fig. 7).

	Tokuyama Bay (mercury contamination) (1)	Léman (eutrophication) (2)
T0	1952	$\cong$ 1950's
T2	?	$\cong$1950–60 scientists and fishermen
TR1		1962 (blooms)
TL1	1964 outbreak of Minamata disease	Not reached
T4	?	1964/66
T5	1967 First effluent standard in Japan	1965–70 sewage collection
TM	1970 max Hg in fish	1976 max P level
T6A	1974 Hg cell ban in industrial process	1975–82 P removal in sewage
T6B	1975/76 Sediment dredging	1985 P ban in detergents
TR2	1983 Fish control lifted	2000–2010? Oligotrophy
Critical period	$\cong$ 20 years	0 year
Policy lag	> 10 years	2 years
Impact duration	22 years	> 35 years
Remediation	9 years	30 years so far
Ecosystem reaction	16 years (limited by dredging)	30 to 40 years

(1): from Nakashini et al (1989) (2): from Rapin et al (1995)

Each documented environmental evolution can be decomposed into the milestones and duration periods as presented here. In table 2, two examples of such decomposition for the Lake Léman eutrophication and for the Tokuyama Bay mercury contamination in Japan are given. In the latter case, environmental reclamation of the bay was accelerated, as the natural recovery process was much too slow. The surficial contaminated sediments were dredged and the fishing ban was lifted after 7 years. Both of these examples refer to chronic and gradual pollution cases. In severe accidental spills the response and recovery of CAS range from relatively short periods for rivers (as for the Sandoz fire spill in Basel on the Rhine in 1986), to decades for groundwater and lakes.

The Future

The future of riverine fluxes and quality, their position in the Earth System and in Human Development depends on three main factors:
(i)   the evolution of river runoff as impacted by waters use,
(ii) the evolution of the contamination of continental aquatic systems, and
(iii) the evolution of river runoff in the context of climate change.

In slowly industrialised regions such as Europe, early impacts can be noted as early as 2,000 years ago (Roman mining and smelting), then developed in the Middle Ages, and accelerated during the Industrial Revolution (see fig. 8).
In developing regions, as in parts of Africa, South America and Asia, environmental governance generally lags behind the development rate, a trend which has been also quite common in industrialised region between 1800 and 1950. When the development rate is very fast policy instruments, their enforcement, and tech-

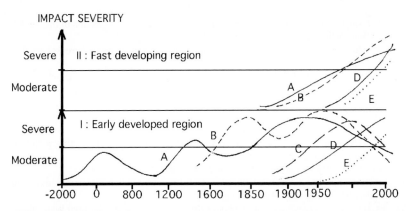

Fig. 8. Hypothetical evolution of water quality issues in early developed (I) and fast developing (II) regions for few key issues. A: metal contamination, B: organic and faecal pollutions, C: eutrophication, D: nitrate pollution, E: organic micropollutants. Note the acceleration of issue occurrence.

nical facilities to control environmental issues concerning CAS, are in most cases poorly equipped to face rapidly emerging and multiple issues. While Europe and North America had 50 to 150 years to cope with water quality issues, the developing countries may have only 10 to 20 years. Unless major efforts are made in environmental assessments – in many countries the levels of issues is not known – and in environmental control adapted to each type of development, these countries will have major water quality issues, which will be enhanced by any reduction of river runoff.

Acknowledgements: many thanks to my colleague Neil Hamilton (IHDP) for editing a draft of this paper. I also wish to express my gratitude to the many colleagues of the UNEP/WHO Gems Water programme, particularly with R. Helmer, D. Chapman and J. Chilton for early discussions on global water quality issues and to the authors of the Water Quality Assessment of the Former Soviet Union, to N.E. Peters (US Geological Survey), to my colleagues of the Piren-Seine programme lead by G de Marsily and G. Billen, to many IGBP scientists, with special mention to C. Vörösmarty, for many fruitful debates on these questions, and to A. Ragu for his continuous and precious collaboration.

References

Billen G, Garnier J, Meybeck M (1998) Les sels nutritifs: l'ouverture des cycles. In: La Seine en son bassin. Fonctionnement écologique d'un système fluvial anthropisé, M. Meybeck, G. de Marsily, E. Fustec (eds) Paris, Elsevier, pp 531–565

Blanc G, Lapaquellerie Y, Maillet N, Anschutz P (1999) A cadmium budget for the Lot-Garonne System (France). Hydrobiol 410: 331–341

Chernogaeva G, Lvov AP, Georgievsky VY (1998) Water use and their influence of anthropogenic activity. In: Kimstach V, Meybeck M, Baroudy E (eds) A water quality assessment of the Former Soviet Union. E.F. Spon London, pp 69–94

Chilton J (1989) Salts in surface and groundwaters. In: Meybeck M, Chapman D, Helmer R (eds). Global Freshwater Quality. A first Assessment. Blackwell Reference London, pp 139–157

Dynesius M, Nilsson C (1994) Fragmentation and flow regulation of river systems in the Northern Third of the world. Science 266: 753–762

Garrels RM, Mackenzie FT (1971) Gregor's denudation of the continents. Nature 231 (5302): 382–383

Horowitz AJ, Meybeck M, Idlafkih Z, Biger E (1999) Variations in trace element geochemistry in the Seine Basin based on floodplain deposits and bed sediments. Hydrological Processes 13: 1329–1340

Humborg C, Ittekot V, Cociasu A, Von Bodungen B (1997) Effect of Danube river dam on Black Sea. Biochemistry and ecosystem structure. Nature 386: 385–388

Leblanc M, Morales JA, Borrego J, Elbaz-Poulichet F (2000) A 4500 years old mining pollution in Spain. Economic Geology, in press

Mainguet M (1999) Aridity Droughts and Human Developments. Springer 302 p

Mariotti A, Letolle R, Blavoux B, Chassaing B (1975) Determination par les teneurs naturelles en 15N de l'origine des nitrates. CR Acad Sci Paris 280: 423–426

Meade RH, Parker RS (1985) Sediments in rivers of the United States. In: National Water Summary 1984, US Geol. Survey Water Supply Paper 2275. US Geol Survey, Reston VA, 49–60

Meybeck M (1979) Concentration des eaux fluviales en éléments majeurs et apports en solution aux océans. Rev Géol Dyn Géogr Phys 21 (3): 215–246

Meybeck M (1982) Carbon, nitrogen and phosphorus transport by world rivers. Amer J Sci 282: 401–450

Meybeck M (1996) River water quality, global ranges tine and space variabilities. Vehr Int Verein Limnol 26: 81–96

Meybeck M (1998) Man and river interface: multiple impacts on water and particulates chemistry illustrated in the Seine river basin. Hydrobiologia 373/374: 1–20

Meybeck M (1999) The IGBP Water Group: a response to a growing global concern. Global Change Newsletter 36: 8–12

Meybeck M, Helmer R, Forstner U, Chilton J (1989) Global waters quality assessment. In: Meybeck M, Chapman D, Helmer R (eds). Global Assessment of Fresh Waters Quality – A First Assessment. Basil Blackwell press, Oxford, pp. 271–292

Meybeck M, Ragu a (1997) Presenting Gems Glori, a compendium of world river discharges to the oceans. Int ass Hydrol Sci Publ 243: 3–14

Nakashini H, Ukita M, Sekine M, Murakami S (1989) Mercury pollution in Tokuyama Bay. In: Sly PG, Hart BT (eds) Sediment/Water Interaction. Kluwer, pp 198–208

Rabalais NN, Turner RE, Justic D, Dortch Q, Wiseman WJ (1999) Characterization of Hypoxia. In: Report for the Integrated Assessment of Hypoxia in the Gulf of Mexico. NOAA Coastal Ocean Program, Decision Analysis Series n°15

Rang MC, Schouten CJJ (1988) Major obstacles to water quality management II Hydro-inertia Verh Int Verein Limnol 23: 1482–1487

Rapin F, Blanc P, Pelletier JP, Balvay G, Gerdeaux D, Corvi C, Perfetta J, Lang C (1995) Impacts humans sur les systèmes lacustres: exemple du Léman. In: Pourriot R, Meybeck M (eds) Limnologie Générale. Masson Paris, pp. 806–840

Schwartz HE, Emel J, Dickens WJ, Rogers P, Thompson J (1990) Water Quality and flows. In: Turner BL (ed) The Earth as Transformed by Human Action. Global and Regional Changes. In: The Biosphere Over the Past 300 years. Cambridge University Press, pp. 253–270

Shiklomanov IA (1998) World Water Resources. A new appraisal and assessment for the 21 ST Century. UNESCO, Paris, 37 pp

Tsirkunov VV, Akuz IK, Zenin AA (1998) The lower Don basin. In: Kimstach V, Meybeck M, Baroudy E (eds) A Water Quality Assessment of the Former Soviet Union. E. and S.N. Spon, London, pp 375–412

Turner RE, Rabalais NN (1991) Changes in Mississippi river water quality this century and implication for coastal food webs. Bioscience: 140–147

Valette-Silver NNJ (1993) The use of sediment cores to reconstruct historical trends in contamination of estuarine and coastal sediments. Estuaries 16: 577–588

Vollenweider RA (1989) Eutrophication. In: Meybeck M, Chapman D, Helmer R, (eds). Global Assessment of Fresh Waters Quality – A First Assessment, Basil Blackwell press, Oxford, pp. 107–120

Vörösmarty CJ, Meybeck M, Fekete B, Sharma K (1997) The potential impact of neo-castorization on sediment transport by the global network of rivers. Int Ass Hydrol Sci Publ 245: 261–282

Walling DE (1999) Linking land use, erosion and sediment yields in rivers basins. Hydrobiologia: 223–240

Zhulidov AV, Emetz VM (1998) Heavy metals, natural variability and anthropogenic impatcs. In: Kimstach V, Meybeck M, Baroudy E (eds) A Water Quality Assessment of the Former Soviet Union. E.F. Spon London, pp 529–558

CHAPTER 7

Integrated Management of Water Resources 7

JAMES C.I. DOOGE*

It is impossible to understand the working of the earth system without a clear understanding of the role of water as part of both the physical geosystem and of the biosphere. It has become increasingly clear that such scientific knowledge must be supplemented by a clear understanding of the interactive relationship between water resources and social systems. To tackle both of these tasks requires new skills and a new receptiveness if the interdisciplinary effort involved is to prove fruitful.

The key role of water in relation to the global environment has been summarised as follows (Ayibotele and Falkenmark 1992):

"(1) Water is a unifying agent of the natural ecosystems with functions similar to the blood and the lymph of the human body.

(2) Water is consumed in bio-mass production which is therefore limited by local water availability.

(3) Water circulation is an important element of the global cycles and in this sense is intimately related to the climate.

(4) Water is a fundamental resource on which depend the life support systems and which has to be equitably shared between all these living in a particular river basin.

(5) Water is a crucial link in the causality chain producing bio-diversity disturbances."

Other experts might add to this list or argue about the balance of the importance of different aspects of the general problem. It is clear however that these questions pose a formidable challenge.

Even to outline the scope of the effort involved in meeting this challenge is more than can be contained in a single presentation of this type. Accordingly, attention will be focussed on three topics from among the many that would be equally relevant. The first is concerned with the unexpected nature of the physical and chemical properties of water which appears to our senses such a simple uncomplicated substance. The next topic under the theme of allocation of water

* e-mail: civil.eng@ucd.ie

use deals with the need for appropriate economic systems and appropriate social systems to enable us to take full advantage of the contribution made to the solution of human problems in this area by science and technology. The third topic chosen discusses how we can cope with the problems of disasters arising from either too much water giving rise to floods or too little water giving rise to droughts.

Understanding the Behaviour of Water

We read from time to time of costly space experiments designed to determine whether water exists or did exist on other planets thus increasing the probability of life in those locations. If an intelligent observer on one of these planets were to examine what we call Planet Earth, he or she (or whatever appropriate gender) would be much more likely to describe it as Planet Water. The first impression would be of the planetary surface covered largely by water in either its liquid or its solid form. A closer examination would reveal a vigorous hydrological cycle involving continuous transformation of water between its liquid, gaseous and solid forms.

The evolution of the present surface temperatures of the planets in the solar system has been a complex one dependent on the evolution of the respective planetary atmospheres. A highly simplified picture would reduce this complex process to a uniform transition from (a) an initial condition of a surface temperature appropriate to a black body at the relevant distance from the sun to (b) an equilibrium condition of surface temperature and pressure based on the formation of an atmosphere by outgassing from the planetary interior (Goodye and Walker 1972). In the case of Mars, which is 228 million kilometres from the sun, the initial temperature would have risen only a few degrees before the water in the atmosphere was in equilibrium with the saturation vapour pressure of ice at about -55 °C resulting in a frozen hydrosphere. In the case of Venus, at the closer distance of 108 million kilometres from the sun with an initial temperature of about 50 °C, the equivalent equilibrium would not be reached and the accumulation of carbon dioxide (CO_2) in the atmosphere of Venus would produce a runaway greenhouse gas effect increasing the surface temperature from 50 °C to over 450 °C and precluding a liquid hydrosphere (Rasool and de Bergh 1970).

In the case of our own planet at an intermediate distance of 150 million kilometres from the sun, the equivalent process would be a moderate greenhouse effect with an increase of mean surface temperature from a few degrees below 0 °C to about 15 °C thus allowing a liquid hydrosphere to develop unlike Mars or Venus. The resulting values of pressure and temperature are sufficiently close to the triple point of water for the variations in space and time to sustain a vigorous hydrological cycle involving all three phases of water. It has been estimated (a) that if the earth were 5 % closer to the sun, there could have been a runaway greenhouse effect resulting in a hydrosphere of water vapour only and (b) that if 5 % further from the sun, equilibrium could have been reached at a temperature below 0° thus producing a frozen hydrosphere (Kondratyev and Hunt 1982).

To the physical chemist, water (H_2O) is a group VI anhydride but its physical and chemical characteristics are quite different to other anhydrides in the group all of which are gaseous at 15°C (Dooge 1983). This anomaly is due to the existence of hydrogen bonding arising from the fact that the water molecule is a polar molecule i.e. is electrically balanced but with a dipole moment. There are other anomalous properties of water that have a direct influence on geophysical processes and consequently an indirect influence on social problems. Since the surface tension of water is anomalously high to a remarkable extent, the retention of liquid water around the contact points between soil particles in an unsaturated soil is greater than it otherwise would be, thus promoting a longer survival for vegetation during drought periods. Among liquids, water is an almost universal solvent and thus provides a key pathway both for necessary nutrients and for toxic substances. The latent heat, the specific heat and the thermal conductivity of water are all anomalously high with the result that the global water cycle acts as an efficient modifier of the variation of climate with latitude and hence of the extent of territory available for comfortable habitation and efficient crop production.

Understanding the complexities of water behaviour is further hampered by the range of scale involved from the water molecule to the global water circulation. To understand the properties and behaviour of water requires knowledge at a molecular scale of 10^{-10}m where water molecules are forming and reforming molecular clusters at a rate of 10^{-13} seconds. To understand the role of water in climate variation and change requires a general circulation model simulating inter-annual variations on a scale of 3×10^6 seconds on a grid scale of the order of 10^5m. Upscaling or downscaling across the complete range of 15 orders of magnitude in space and 20 orders of magnitude in time remains a distant dream. It is little wonder that 400 years ago Galileo said:

"I can learn more of the movement of Jupiter's satellites than I can of the flow of a stream of water."

or that 250 years ago Euler, who established hydrodynamics as a science, wrote:

"What authors have written up to now on the motion of rivers is of little moment; and all they have said is founded only on hypotheses which are arbitrary, even altogether false ... But this problem is so difficult that, though it depends only on analysis, we can almost never hope to come to the general solution, which could serve to determine the motion of every kind of river."

Since then hydraulics has succeeded in extending understanding of the movement of water at the scale of the continuum point (10^{-5} m) and the turbulent mixing length (10^{-2} m) and hydrology has made advances at the scale of the experimental plot (1 m) and at sub-basin (10^3 m) and basin scale (Dooge 1986). Much remains to be done both in relation to behaviour at given scales of interest and on linkages between scales.

The hydrological cycle itself interacts with other geophysical cycles and with complex economic and social systems. The hydrological cycle interacts strongly with the cycle of erosion and sedimentation and with the main bio-geochemical cycles. When we turn from the consideration of hydrology as a component of geophysics to a consideration of the role of hydrology in helping to solve social problems, the situation is again complex. The combination of these geophysical cycles affects the natural environment which is subject to economic development and population growth. As a result there are changes in both water resource use and the nature of the geosystem which provide a feedback into the basic cycles of water, sediment, chemicals and living organisms.

Allocation of Water Use

The growth of population and the pace of economic development have combined to produce an enormous increase in water needs over the past 100 years as shown by table 1 (based on Shiklomanov 1991).

Agriculture remains the largest consumer of water but the proportional increase in demand from other sectors is much greater. There is an enormous variation in the available water per head of population from region to region and from country to country. The surplus available water varies from 300 cubic metres per year per head in the Arabian Peninsula to 170,000 cubic metres per year per head in Canada and Alaska (Shiklomanov 1997). It is little wonder that the 1991 Dublin International Conference on Water and the Environment (ICWE) stated in the first of its four guiding principles (Young et al. 1994, p 162)

"*Water is a finite and vulnerable resource, essential to sustain life, development and the environment.*"

and went on to stress:

"*Since water sustains life, effective management of water resources demands a holistic approach, linking social and economic development with protection of natural ecosystems.*"

In the remainder of this section, attention is concentrated on the economic and social structures needed to take full advantage of our scientific and technological knowledge in the field of water.

Table 1. Water needs by use

Type of use	WATER USE (km³/year)		
	1900	1950	2000
Agriculture	525	1130	3250
Industry	37	178	1280
Municipal	16	52	441
Reservoir LOSS	0.3	6.5	220
Total	579	1366	5190

The economic context has been well discussed by the Business Council for Sustainable Development, a group of prominent leaders of international industrial companies who provided an input to the Rio Conference. Their report clearly states (Schmidheiny 1992, p 16):

"The market does not tell us where to go, but it provides the most efficient means of getting there."

and goes on to draw the conclusion (Schmidheiny 1992, p 17):

"Therefore society – through its political systems – will have to make value judgements, set long term objectives, implement measures such as charges and taxes step by step and make mid-course corrections based on experience and changing evidence."

This important distinction between the irrelevance of market forces in setting objectives and the efficiency of market forces in achieving agreed goals is too often blurred or ignored in practice with consequent short term gains for a few at the cost of a long term burden on the community as a whole.

In addition to appropriate economic systems, we need appropriate social systems in dealing with water resource problems. The second of the four guiding principles of the Dublin ICWE Conference deals with this point (Young et al. 1994, p 17):

"Water development and management should be based on a participatory approach. involving users, planners and policy-makers at all levels."

and continues by outlining the elements of such an approach:

"The participatory approach involves raising awareness of the importance of water among policy-makers and the general public. It means that decisions are taken at the lowest appropriate level, with full public consultation and involvement of users in the planning and implementation of water projects."

Such a participatory approach called for new skills in communication and a recognition that true communication is a two-way process.

In an informative report on rural water supplies in Lesotho by a multidisciplinary team (Feachem et al. 1978), an interesting model is described aimed at avoiding the dangers of over-centralisation on the one hand and over-decentralisation on the other in the successive phases of a rural water supply. The authors refer to this compromise model which is set out in table 2 as controlled self-help.

A key to successful local participation in all schemes for development is the involvement of the local community at the earliest possible moment in the planning phase. Equally important is the requirement that national and local governments should maintain an interest after the completion of the initial scheme. The same interdisciplinary team mentioned above – which included experts in envi-

Table 2. Controlled self help

Activity	Central Authority	District office	Local Community
Establish	Select criteria	-	-
Plan	Gather information Allocate resources to districts	Select villages Invite applications	Preliminary organisation Respond to invitation
Fund	Disburse grants Distribute donor funds	Collect local contributions Prepare project documents	Raise funds and labour on agreed basis
Implement	Technical advice Central purchasing	Technical supervision	Contribute labour
Maintain	Provide funds	Employ staff	Volunteer Labour or services

ronmental engineering, community health, economics, social anthropology, administration and management – stressed the need for complementary inputs to obtain the maximum return throughout the lifetime of the project for the original investment of financial and human resources (Feachem et al. 1978). They stressed that the immediate benefits of improved health and reduction of labour in water-carrying cleared the way for full economic rural development in successive stages as indicated on table 3.

Table 3. Complementary inputs for full benefit

Aim/Benefit	Complementary Input
Immediate	Community participation Competent design Adequate maintenance
Stage I	Use of new facility New pattern of water use New habits of hygiene No new hazard

Stage II	Adequate service for
	   agricultural advice
	   training
	   marketing
	   credit

Stage III	Integrated rural development programme

Coping with Water-related Disasters

Water so essential to life, health and well-being can also be a source of danger. Pliny the Elder (Gaius Plinius Secundus in 77AD) related the social impact of the Nile Flood to the nilometer readings in the following fashion. A height of 14 ells resulted in happiness, 15 ells in security, and 18 ells in disaster due to flooding; below the happiness level of 14 ells, 13 ells corresponded to suffering and 12 ells to widespread hunger. Two millenia later, the world is still faced with disasters arising from the extremes of too much water or too little water.

There are many types of water-related disasters. In flooding, there are distinct differences between the characteristics and the consequences of (a) flash floods in upland areas, (b) combinations of tributary flows producing flooding over a large area of the middle catchment, and (c) perennial lowland flooding near the estuary of a large river. Droughts also differ in their characteristics and their consequences between arid areas and semi-arid areas. In arid areas the shortage of water is the norm and the number of inhabitants and their life style are in approximate balance with the harsh environment. In semi-arid areas, the occurrence of a wetter than average season, or a short sequence of such seasons, encourages an increase in livestock numbers and consequently both increased losses and serious land degradation in a subsequent dry period. There are also other water-related disasters that are not directly related to the total amount of water in the system. These include water-induced instabilities such as glacier outflows, mud flows and rock slides, storm surges due to wind-water interaction, and major incidents of water pollution.

Among the people affected by natural disasters the vast majority are affected by flood or droughts, though the resulting deaths are greater in the case of earthquakes and windstorms. Figures for the period of 1960 to 1979 due to Hagman (1984) are shown on table 4.

More recent figures for sudden natural disasters (thus excluding droughts) in 1997 indicate that flooding accounted for 26 % of all events, 43 % of deaths and 46 % of economic losses (Munich-Re 1998).

In dealing with disasters it is necessary to distinguish between the causal hazard (geophysical or technological or man-made) and the impact of this incident on a given group of people. The link between the hazard and the disaster is the

Table 4. Impact of natural disasters (1960–1979)

Type of disaster	Percentage of events	Percentage of persons affected	Percentage of persons killed
Floods	30	29	5
Droughts	10	60	18
Windstorms	24	8	33
Earthquakes	14	2	33
Others	22	1	11
Total	100	100	100

vulnerability of those impacted. Blaikie et al (1994) defined this concept as follows:

> "By vulnerability we mean the characteristics of a person or group in terms of their capacity to anticipate, cope with, resist, and recover from the impact of the natural hazard."

It is being increasingly realised that coping with natural disasters depends largely on our ability to mitigate the vulnerability of the affected groups through improved risk assessment and planned disaster preparedness.

Despite the variety of phenomena involved in coping with the various types of hazards, all water-related disasters (and indeed all types of disasters) involve the same sequence of processes. These may be grouped in five phases: (1) the anticipatory phase, (2) the alarm phase, (3) the impact phase, (4) the relief phase, and (5) the rehabilitation phase. The anticipatory phase includes the mapping of risk assessment based on hazard probability and local vulnerability, establishment of codes of practice, design of structural measures, planning control and insurance, training and information programmes. The warning phase includes continuous monitoring of hazard indicators by experts, advanced warning system for local services and finally a public alert. The impact phase, which depends on local response, includes search and rescue efforts, emergency repairs of services and evacuation. The relief phase, which involves external help, is concerned with emergency supplies, full restoration of services, temporary housing, and so on. The final rehabilitation phase, which may be prolonged, includes physical reconstruction, personal rehabilitation and post-hoc evaluation of the efficiency of the procedures under each of the five phases.

The realisation that mitigation of vulnerability offers the best hope of reducing the impact of disasters poses a number of questions in relation to overall policy formulation. It is becoming clearer that money spent now on disaster mitigation can both reduce human misery and financial expenditure in the long run. This raises ethical as well as economic questions. The Commission of Ethics in Science and Technology of UNESCO (COMEST) has established a Working Group on Ethics in relation to Fresh Water which has been considering a number of aspects of this general problem including the ethics of water-related disasters.

It is a long but interesting road from the properties of water molecules to the ethics of freshwater management. Each of the elements along the way is important in understanding the occurrence and movement of water on planet earth and determining the optimum way to manage this resource so necessary for life, health, and economic development. Success depends on the establishment of real partnerships between a number of disparate groups and the development of new procedures for achieving consensus and implementing the agreed programme. The task is a formidable one but the rewards would be great.

References

Ayibotele NB, Falkenmark M (1992) Freshwater Resources. In: Dooge JCI, Goodman GT, La Riviere JWM, Marton-Lefèvre J, O'Riordan T, Praderie F (eds) An Agenda of Science for Environment and Development into the 21st Century. Cambridge University Press, Cambridge pp 187–203

Blaikie O, Cannon T, Davis I, Wisner B (1994) At Risk. Natural Hazards. People Vulnerability and Disasters, Routledge

Dooge JCI (1983) On the study of water. Hydrological Sciences Journal 28 (1): 23–48

Dooge JCI (1986) Scale Problems in Hydrology. Kisiel Memorial Lecture, February 1986, University of Arizona. Reproduced in: Buras N (ed) (1997) Reflections on Hydrology. American Geophysical Union, Washington, pp 84–145

Dooge JCI (1995) Searching for simplicity in hydrology. Surveys in Geophysics 18: 511–534

Feachem Rg and 8 co-authors (1978) Water, Health and Development. Tri-med Books, London

Golubev GN (1983) Economic activity and water resources in the environment: a challenge for hydrology. Hydrological Sciences Journal 28 (1): 57–75

Goodye RN, Walker JCG (1972) Atmospheres. Prentice-Hall

Hagman G (1984) Prevention Better than Cure, Swedish Red Cross. Stockholm and Geneva

Kondratyev KY, Hunt GE (1982) Weather and Climate on Planets. Pergamon Press, Oxford

Munich-Re (1998) Annual Review of Natural Catastrophes 1997. Münchener Rückversicherungs-Gesellschaft, Munich

Rasool SI, de Bergh C (1970) The runaway greenhouse and accumulation of CO2 in the atmosphere. Nature 226: 1037–1039

Schmidheiny S (1992) Changing Course. A Global Business Perspective on Development and the Environment. M.I.T. Press, Cambridge

Shiklomanov I (1991) The World's Water Resources. In: International Symposium to commemorate 25 years of IHD/IHP. UNESCO, Paris, pp 93–126

Shiklomanov I (1997) Assessment of Water Resources and Water Availability of the World. Comprehensive Freshwater Assessment. World Meteorological Organisation, Geneva

Young GJ, Dooge JCI, and Rodda J (1994) Global Water Resource Issues. Cambridge University Press, Cambridge

CHAPTER 8

The Challenge of Global Water Management   8

CARLO C. JAEGER*

Rising standards of living and growing human population make scarcity of freshwater resources an increasingly serious problem, which can amplify destructive and cruel social conflicts. World-regions where this problem is most relevant include large parts of Africa, the Middle East, China and India.

By far the largest fraction of global freshwater use is due to evaporation in the course of food production. Within the next decades, this quantity is bound to increase due to rising incomes, changing diets, and population growth. This will generate strong pressures to use water for food production more effectively. One likely effect will be massive trade in "virtual water", i.e. food trade induced by water scarcity. Arid and semi-arid regions in Africa and Asia are faced with the problem of building up export capacities to finance increasing food imports in the future and to generate badly needed employment opportunities which cannot be provided by agriculture.

It is instructive to design a global institutional arrangement which would provide the single best solution to this problem if such a solution were available. As this is not the case in reality, however, researchers and policy-makers should engage in a patient and careful search for viable second-best solutions. This will involve new definitions of property rights for various kinds of water use. Moreover, building up export capacities in water scarce regions could require something like a global Marshall plan – this time with Europe at the giving rather than at the receiving end.

Water Resources and Social Conflict

In the 1990ies, a debate about water resources and social conflict has been triggered by several authors (Postel 1992, Gleick 1993, Shiklomanov 1993, Clarke 1994, Falkenmark 1997). They showed that an increasing number of regions and countries was bound to experience serious water stress in the foreseeable future. Their arguments were reinforced by work at the World Bank on Water Resources Management (World Bank 1993, Serageldin 1995). In close relationship with this debate, Homer-Dixon (1991) and Lonergan and Kavanaugh (1991) emphasised the relevance of water stress for social conflict and environmental security. This

---

* e-mail: carlo.jaeger@pik-potsdam.de

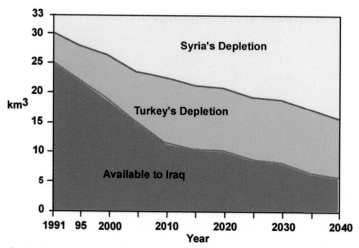

Fig. 1. Projected sequential depletion of the Euphrates river. [Source: Kolars and Mitchell (1991).]

was also highlighted in the comprehensive discussion of today's freshwater problems given by WBGU (1999). A case in point is the Euphrates river, as illustrated in figure 1.

One should not necessarily expect conflicts of interest over water resources to become explicitly addressed in the form of open social conflict. It is much more likely that conflicts which are fought in the name of other issues become more intense and intractable if they are mixed up with conflicts over water resources. As long as the parties involved enjoy a considerable level of welfare and share procedures for orderly negotiations, management of water resources need not lead to major conflicts. An example for rather successful management of a river flowing through several countries fulfilling these conditions is given by the river Rhine (Bernauer 1997). On the other hand, it is obvious that most countries of the world do not share these favourable conditions.

Less favorable conditions are usually characterised by attempts to raise low standards of living which risk to jeopardise the stock of environmental resources. Such seems to be the case in Southern Africa, as figure 2 illustrates.

A crucial step in the debate about water resources and social conflict was made with the introduction of the notion of virtual water (Raskin et al. 1995, Allan 1996). It refers to the water which is necessary to produce some commodity. The concept is especially relevant for food trade, as many countries have found ways of alleviating water stress by importing food – they are said to be importing a corresponding amount of virtual water. And of course this sort of food trade becomes increasingly important as differences in water scarcity increase at a global scale (Alcamo et al. 1997).

Meanwhile, the challenge of global water management has led to comprehensive initiatives towards water assessment (e.g. UNCSD 1997 and the related back-

Fig. 2. A framework for the analysis of sustainable development in Southern Africa. Source: Turton (1999).

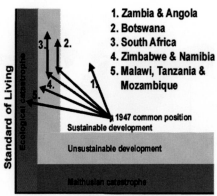

1. Zambia & Angola
2. Botswana
3. South Africa
4. Zimbabwe & Namibia
5. Malawi, Tanzania & Mozambique

1947 common position
Sustainable development

Unsustainable development

Malthusian catastrophe

Standard of Living

Ecological catastrophe

Stock of Environmental Resources

ground papers) and first attempts to identify future developments of water regimes (Brunnée and Toope 1997, Rosegrant 1997).

Food and Water

Projections of food security suggest that water availability will be a key factor in the decades to come. Modelling global freshwater use is a challenging task (Hoekstra 1997), but there can be little doubt that serious regional imbalances in freshwater availability will increasingly shape global agricultural dynamics in the next decades.

Water availability, soil characteristics and climate conditions make North America and Europe – although with important internal regional differences – world regions which will be able to export major amounts of food in the decades to come. The same holds for parts of Latin America, Australia and some South-East Asian regions. On the other hand, India and China may be able to satisfy their demand for food from internal supply only if they accept that their main rivers don't reach the oceans anymore and if they manage to redress major internal imbalances in regional water availability. Finally, while considerable parts of Asia and Africa may be able to feed themselves without becoming food exporters, large parts of the Middle East and of Africa are likely to become major food importers because of water shortage (see fig. 3).

The resulting outlook is not necessarily grim, but it certainly presents a major challenge. To appreciate the dimension of this challenge, it is important to acknowledge that the largest quantity of water used by humankind is used for agricultural purposes (see fig. 4).

There are two ways to define agricultural water use, and the difference is quite important. On the one hand, one may look at water used for irrigation. One will then include water which could be reused, and one will neglect evapotranspiration from rainfall and soil moisture which obtains independently from irrigation. Present estimates for water use in irrigation lie around 3,500 km^3 of water per

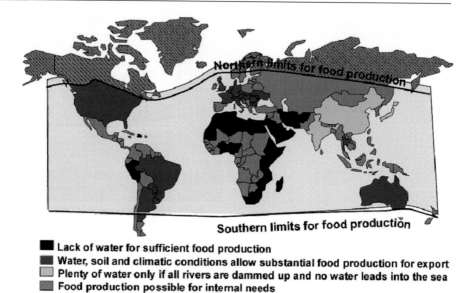

Lack of water for sufficient food production
Water, soil and climatic conditions allow substantial food production for export
Plenty of water only if all rivers are dammed up and no water leads into the sea
Food production possible for internal needs

Fig. 3. Water availability and food security. [Source: Zehnder et al. (1999).]

year (Gleick 1993). Reducing the irrigation water needed to produce 1 kg of plant material by half is perfectly feasible in many cases. However, in order to increase global food production the total amount of irrigated land will strongly increase. So far, this trend is stronger than increasing irrigation efficiency.

The other definition of agricultural water use refers to total evaporation from agricultural production, including evapotranspiration from plants and evapora-

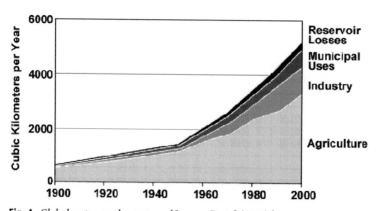

Fig. 4. Global water use by sectors. [Source: Postel (1992).]

tion from irrigation water (Zehnder 1997). As a rule of thumb, it takes about 1 m^3 of water to produce 2 kg of dry wheat plant (Cohen 1996). This in turn yields 1 kg of bread or about 2,500 kilocalories of food. The numbers for different varieties of cereals and other edible plants vary; however, the order of magnitude is fairly constant. As for meat, it takes more than 5 kg of plant material to produce 1 kg of meat. Accordingly, the amount of water needed to produce 1 kg of meat is much greater than for vegetable food. The efficiency of evapotranspiration may be increased by breeding – possibly with the support of genetic engineering – suitable crops and animals. However, the physiological processes involved seem to set rather narrow limits to such efforts (Muller 1974). While productivity per square meter of soil can be greatly enhanced, the biochemistry of photosynthesis is such that productivity per cubic meter of water is hard to increase beyond a few percentage points.

Why should we care about evaporation as long as the water is not coming from irrigation? Because the water so evaporated will not run off any more via rivers, and may no more be available for human use. Precipitation is a renewable, but also a limited resource: when it is used up, this makes a difference, whether the use is based on irrigation or not. In principle, evaporation may result in additional rainfall available on contiguous agricultural land. This gives some leeway in the restrictions set by precipitation quantities. Therefore, detailed research about such situations is warranted. However, the recycling of precipitation via clouds does not turn limited freshwater resources into unlimited ones.

If we consider only evapotranspiration, a vegetarian diet of 2,500 kilocalories per capita and day implies a water use of about 1 m^3 of water per capita and day or about 350 m^3 of water per capita and year. A mixed diet including 20 % calorie uptake from meat will double this requirement up to about 700 m^3. These are theoretical lower bounds, in practice the water requirement will be much higher due to pre- and post-harvest losses and to evaporation from irrigation. An extremely conservative estimate may set total evaporation for a mixed diet at about 1,500 m^3 of water per capita and year.

Presently, meat provides only about 5 % of total calorie uptake as a global average, and total population is 6 billion people. Under these conditions, total evaporation from food production amounts to about 1,000 m^3 of water per capita and year, leading to a total of about 6,000 km^3 of water per year. These are huge figures, and they do not yet include water use for industry, services, and households.

Looking at the future, the meat component in human diet may well increase towards about 20 % calorie uptake from meat, and human population is likely to increase to about 10 billion people. With these estimates, food production around the year 2050 would use about 15,000 km^3 of water per year. Together with other forms of human water use, we may reach as much as 20,000 km^3 of water per year.

Virtual Water

This is a sizeable fraction of total precipitation over continents and islands, estimated at about 100,000 km^3 of water (Postel 1992, Cohen 1996). About two thirds of this total, or 65,000 km^3, evaporate from the ground, from water surfaces or via plants.

The remaining 35,000 km^3 run off via rivers. However, about 25,000 km^3 run off under conditions which make it very hard to utilise the water in question. This holds for remote areas, but also for sudden runoffs after heavy rains and the like. Of course, agricultural evapotranspiration is not based only on irrigation from managed run-off, but the difference of 10,000 km^3 is by no means huge if compared to the water requirements which are to be expected for future food production. This already hints at the serious issue of large rivers being depleted to the point where they do not reach the ocean any more.

One may imagine that at the global level evaporation from precipitation is large enough to take care of the water needs for global food production. Of course, present agricultural evapotranspiration of about 5,000 km^3 is included in the figure for total evaporation indicated above. And if agricultural evapotranspiration increases, this will to some extent replace previous evapotranspiration from non-agricultural plants. However, increasing food production implies increasing net biomass production and thereby increasing total evapotranspiration. This will lead to some additional precipitation via cloud formation. But there is no way for this effect to provide all the additional water required.

Of course, there are many ways to increase the amount of water available for human use. They include genetical engineering plants which are able to use salt water, or attempts to catch precipitation over the oceans, either in the form of icebergs or of huge plastic vessels filled by rainfall, and bring them into places with water shortage. All these options involve considerable economic costs.

While in many production processes increasing returns to scale lower the costs if quantities increase, in the case of water supply it is more likely for costs to rise because of resource scarcities. Desalination, for example, requires so much energy that large scale desalination schemes could drive energy prices up (Howe 1974). And it is obvious that large scale attempts to gain energy from fuel crops would complicate the problem of water scarcity even further.

The relevance of costs means that economic mechanisms forcefully enter the picture. Freshwater is a renewable resource in limited supply. Where it can be appropriated, the resource generates a return, and there may be struggles about who gets that income. The amount of freshwater available for human use can be increased to some extent, but this comes at a cost. The problem is greatly exacerbated by the fact that precipitation falls very unequally over different parts of the world (see fig. 5).

Combined with the fact that in the contemporary world economy purchasing power is very unequally distributed, too, this situation generates considerable potentials for social conflict. When trying to deal with this potential, it is essential to ask which comparative advantages in dealing with freshwater resources arise in which regions and countries.

Fig. 5. Annual precipitation in different places. [Source: Unesco (1978).]

Lima (Peru)	**41 mm**
El Golea (Algeria)	**48 mm**
Almeria (Spain)	**218 mm**
Jerusalem (Israel)	**500 mm**
New York (USA)	**1,074 mm**
Tabing (Indonesia)	**4,455 mm**
Cherrapunji (India)	**11,477 mm**

In principle, it is conceivable to grow large quantities of food in, say, sub-Saharan Africa with water deviated from the Congo. However, it is hard to see how such agriculture could compete with food production relying on the abundant rainfall over Europe. In principle, India can expand food production to the extent that its major rivers no longer reach the ocean: but it is hard to see why this should be wiser than importing food from regions with more advantageous relations between precipitation and population. This is not to say that countries faced with water scarcity should simply give up agriculture. In many cases there may be a comparative advantage in developing forms of agriculture which have strong synergies with tourism and other activities and deliberately expanding food imports to alleviate water stress.

Today, about ten percent of global grain production is traded internationally, in the case of soy the fraction is about one quarter. The increasing scarcity and unequal distribution of freshwater resources are powerful factors pushing these proportions up in the decades to come. Actual water shortage can and often will be alleviated by virtual water trade.

Challenges for Research and Policy

Water induced food trade raises a series of difficult challenges. How can importing countries establish export sectors with sufficient strength to finance food imports? Besides generating the currency required for that purpose, growing export sectors are also needed to generate employment for the people who cannot make a living in agriculture. To establish such export sectors, however, requires substantial investments in order to accumulate the necessary fixed capital in terms of infrastructure, equipment, software, etc. Such investments are even more important as the skills and culture needed for a successful export business only thrive in interaction with suitable fixed capital. To prepare for food imports tomorrow, capital needs to be imported today, first as financial capital and then as equipment financed by it.

The problems involved in managing global freshwater resources may be as complex as the ones which were met after World War II with the Marshall plan for Western Europe (Grose 1997). This time, however, Europe finds itself at the giving end of the process – with all the potential advantages but also with all the responsibilities which this implies.

Is there a unique optimal solution to the problems of global freshwater resources, a solution which could be identified by research and implemented by policy-makers? Such would be the case if the global water cycle could be managed as a global commons by some international authority and if the complex system of relevant markets – for drinking water, irrigation water, food, land, capital equipment, financial capital, etc. – exhibited the kind of perfection discussed in micro-economics textbooks (for a thoughtful exposition see Kreps 1990). Under these conditions, it would be sufficient to define appropriate property rights for freshwater and leave the rest to the markets.

It is useful to briefly analyze this hypothetical situation. Suppose that nobody on earth would be allowed to use freshwater resources without first buying a permit to do so. Permits would be available from local authorities or directly via internet from a global water agency, from which local authorities would purchase their permits, too. The permits could be freely traded at whatever price the market would bring about. They would refer to a specific time period and would need to be renewed after that period. The global water agency would decide on the basis of scientific evidence what amount of freshwater resources to make available and issue a corresponding number of permits. Of course, in co-operation with local authorities it would continuously monitor the use of freshwater resources and police users who failed to buy the required permits. The receipts from the sale of permits would be used to compensate those consumers who under the new scheme would be faced by increased costs due to water pricing. As a result, they would be better off than before while still having a clear incentive to reduce demand for water intensive goods and services.

This is the optimal solution to the problem of using a renewable resource with fixed supply in the institutional setting of a market economy. It would result in improved irrigation techniques, in enhanced run-off management, reduced water pollution, and, yes, in increased food trade. Capital markets would make sure that the investments required for efficient global water use took place and humankind would live happily thereafter.

This scheme provides what the sociologist Max Weber called an ideal type, a conceptual tool which highlights important features of a complex social situation by identifying a conceivable, normatively coherent social choice in that situation (Weber 1949). In the present case, four points deserve special attention. First, it is not very likely that a global water authority would use its receipts to compensate the ones most threatened by increasing water prices, namely the poorest. Second, it is not very likely that the nations of the world would surrender their sovereignty in dealing with such a fundamental resource as water to a global central authority. Third, the amount of global freshwater resources to be made available in any period could not be defined on the basis of scientific evidence alone, as very different figures could be defended with sensible arguments.

Finally, actual capital markets have no way of identifying an optimal pattern of investment in view of what the future holds. In today's world economy, futures markets exist for a few selected goods and services and for quite a limited time span, but even this is the exception. Therefore, investment decisions today are not driven by market prices for futures contracts but by the expectations entertained by investors (for a model of this kind of situations see Morishima 1992). In some cases, these expectations may function as self-fulfilling prophecies, in other ones they give rise to a world-wide lottery about unknown future developments. In any case, the expectations of investors are necessarily shaped to a considerable extent by group think among investors, by signals from policy makers, by knowledge claims from scientific communities, etc.

These four points imply that no single best solution is available to deal with the global water challenge of the coming decades. Therefore, the danger of destructive and cruel social conflicts intensified by scarcity of water resources is a very real one. To avoid this danger will require a patient search for second best solutions. The search will not be easy, because so far the promises of a "general theory of second best" have failed to materialise (Lipsey and Lancaster 1956, Srinivasan 1996).

The search for second best solutions to the global water challenge must draw on thorough disciplinary work on specific issues. Given the present argument, scientific studies of evapotranspiration under varying forms of land use are especially important, and so are studies of extreme precipitation events and of the risks which their possibility engenders. On the social side, an often neglected aspect concerns the history of institutions dealing with water. Without differentiated knowledge about these institutions, their legal forms, cultural meanings, past crises, etc., future water policies will be groping in the dark. Technical and managerial knowledge is equally indispensable. Many options must be pursued for quite some time until reliable information about their feasibility as well as their costs and benefits can be obtained.

As long as costs and benefits can be aggregated across all relevant stakeholders, relatively simple decision procedures can be applied. In freshwater assessments, this is likely to be the exception. As a rule, it will be indispensable to assess the pros and cons of different options from the point of view of different stakeholders. For this purpose, truly integrated assessments are often warranted (Rotmans and van Asselt 1996, Jaeger 1998, Pahl-Wostl et al. 1998). What is required is an array of assessments run as communication processes which involve researchers from widely varying disciplines along with major stakeholders, including public authorities, multinational firms, farmers, citizens, and other interested parties. This leads to a further research challenge: developing families of integrated models which take into account the diversity of perspectives involved in the very systems under study.

Global integrated freshwater assessments will become increasingly important in the years to come. When designing such assessments, a plurality of approaches is warranted. There are many sensible ways of studying the earth system in general (Schellnhuber and Wenzel 1998) and in view of its freshwater resources in particular. A modular structure seems especially promising: several research teams deliberately share certain modules – be it software, data, conceptual models, analysis tools, stakeholder links, etc. – while varying other ones. At the same time, they jointly develop procedures to foster comparability.

Of course, it will be essential not only to study problems and design possible measures, but to assess again and again the consequences of those steps which are actually taken (including decisions to just wait and see). Along these lines, it should be possible to generate knowledge and action leading towards a sound second-best solution to the challenge of managing the freshwater resources of the earth.

Acknowledgements

Thanks to A.J.B. Zehnder, R. Schertenleib, and Yang Hong, all of EAWAG, for intense discussions and shared work on global water issues. Thanks to Monika Kaempfer and Ariane Maniglia, also of EAWAG, for designing the figures. I am grateful for the extremely helpful comments to previous versions of the paper obtained at the "Third open conference of the human dimensions of global environmental change research community", Shonan Village, Japan, June 1999, at the BAHC-Workshop on "Freshwater Resources in Sub-Saharan Africa"; UNEP, Nairobi, October 1999, and in a discussion on the "Visions"-Project at ICIS, Maastricht, in February 2000. Thanks for fruitful conversations about topics addressed in the paper go to Axel Bronstert, Bill Clark, Ottmar Edenhofer, Michael Fosberg, Stephan Kempe, Jacques Léonardi, Scott Moss, Jan Rotmans, and John Schellnhuber. Responsibility for errors remains with the author.

References

Alcamo J et al. (1997) Global Change and Global Scenarios of Water Use and Availability: An Application of Water GAP 1.0. Wissenschaftliches Zentrum für Umweltsystemforschung, Universität Gesamthochschule Kassel, Kassel

Allan JA (1996) Water, Peace and the Middle East. Negotiating water in the Jordan Basin. Tauris Academic Studies, London

Bernauer T (1997) Managing International Rivers. In: Oran R, Young (eds) Global Governance: Drawing Insights from the Environmental Experience. MIT Press, Cambridge, Massachusetts

Brunnée J, Toope SJ (1997) Environmental Security and Freshwater Resources: Ecosystem Regime Building. American Journal of International Law 91: 26–59

Clarke R (1994) Wasser. R. Piper GmbH & Co. KG, München

Cohen JE (1996) How many people can the earth support, W.W. Norton & Company, London and New York

UNCSD (1997) Comprehensive Assessment of the Freshwater Resources of the World United Nations Document E/CN.17/1997/9, New York

Falkenmark M (1997) 'Water Scarcity – Challenges for the Future' in EHP Brans, EJ de Haan, A Nollkaemper, J Rinzema (eds) The Scarcity of Water. Emerging Legal and Policy Responses. Kluwer, London

Gleick PH (ed) (1993) Water in Crisis. A Guide to the World's Fresh Water Resources. Oxford University Press, Oxford

Grose P (ed) (1997) The Marshall Plan and Its Legacy, Foreign Affairs

Hoekstra AY (1997) The Water Submodel: AQUA. In: J Rotmans, B de Vries (eds), Perspectives on Global Change. The TARGETS Approach. Cambridge University Press, Cambridge

Homer-Dixon TF (1991) On the threshold: Environmental changes as causes of acute conflict. International Security 16: 76–116

Howe ED (1974) Fundamentals of water desalination. Marcel Dekker Inc., New York

Jaeger CC (1998) Risk Management and Integrated Assessment. Environmental Modeling and Assessment, pp 211–225

Kolars JF, Mitchell WA (1991) The Euphrates River and the Southeast Anatolia Development Project. Southern Illinois University Press, Carbondale

Kreps DM (1990) A Course in Microeconomic Theory. Harvester Wheatsheaf, New York, London, Toronto, Sidney, Tokyo

Lipsey RG, Lancaster KJ (1956) The general theory of second best. Review of Economic Studies 24: 11–33

Lonergan SC, Kavanaugh B (1991) Climate change, water resources and security in the Middle East. Global Environmental Change 1: 272–290

Morishima M (1992) Capital & Credit. A new formulation of general equilibrium theory. Cambridge University Press, Cambridge

Muller WH (1974) Botany: A Functional Approach. Macmillan, New York

Pahl-Wostl C et al. (1998) Integrated Assessment of Climate Change and the Problem of Indeterminacy. In: P Cebon, U Dahinden, H Davies, DM Imboden, Jaeger CC (eds) Views from the Alps. Towards Regional Assessments of Climate Change. Cambridge Unioverity Press, Cambridge, Massuchsetts

Postel S (1992) Last Oasis. W.W. Norton & Co., London & New York

Raskin P et al. (1995) Water and sustainability: a global outlook. Stockholm Environment Institute, Stockholm

Rosegrant M (1997) Water resources in the twenty-first century: Challenges and Implications for Action, Food, Agriculture, and the Environment. Discussion Paper 20. International Food Policy Research Institute, Washington, DC

Rotmans J, van Asselt MBA (1996) Integrated Assessment: A Growing Child on its Way to Maturity. Climatic Change 34: 327–336

Schellnhuber H-J, Wenzel V (1998) Earth system analysis: Integrating science for sustainability. Springer, New York

Serageldin I (1995) Toward sustainable management of water resources. The World Bank, Washington, DC

Shiklomanov IA (1993) World water resources. In: Gleick PH (ed) Water in Crisis. Oxford University Press, New York & Oxford

Srinivasan TN (1996) The Generalized Theory of Distortions and Welfare: Two Decades Later. In: Feenstra R, Grossman G, Irwin D (eds) The Political Economy of Trade Policy: Essays in Honor of Jagdish Bhagwati. MIT Press, Cambridge, Massechusetts

Turton AR (1999) Southern African Hydropolitics: Development Trajectories of Zambezi Basin States and South Africa, Occasional Paper No. 7. Water Issues Study Group. SOAS University of London, London

Unesco (1978) World Water Balance and Water Resources of the Earth. Unesco, Paris

WBGU – German Advisory Council on Global Change (1999) World in Transition: Ways Towards Sustainable Management of Freshwater Resources. Springer, Berlin

Weber M (1949) The Methodology of the Social Sciences. The Free Press, New York

World Bank (1993) Water Resources Management. The World Bank, Washington DC

Zehnder AJB (1997) Wasser: ein knappes Gut? EAWAG news, pp 3–4

Zehnder AJB et al. (1999) Le défi de l'eau. GWA

PART III

Perspective: III
Advancing our Understanding:
Reductionist and/or Integrationist Approaches
to Earth System Analysis

Understanding Climate Variability: A Pre-requisite for Predictions and Climate Change Detection

HARTMUT GRASSL*

It is interesting to note that the person who has worked over the last five years to strengthen and – if needed – to initiate the joint action of the four major Global Change Research Programmes was asked to speak at the conference on "The Disciplinary Way". Maybe because without a high profile in disciplinary research all interdisciplinary and trans-disciplinary research loses its basis and becomes wishful thinking. The content of this contribution is partly a review of the achievements of the World Climate Research Programme (WCRP), the oldest and really globally co-ordinated environmental research programme that attracts the full scientific community. Based on a fore-runner, called Global Atmospheric Research Programme (GARP), and the infrastructure of National Meteorological and Hydrological Services (NMHSs) of about 185 members of the World Meteorological Organisation (WMO) and jointly sponsored since 1980 by WMO and the International Council of Scientific Unions (ICSU, now renamed to International Council for Science), WCRP made major steps forward to reach its major goal: To understand and to predict – as far as possible – climate variability and climate change including human influences.

Causes of Climate Variability

The components of the climate system, namely air, water, life and solid earth, strongly interact at drastically different reaction time scales from hours (planetary boundary layer) to billions of years (complete rearrangement and formation of continents). Thus, there must be climate variability on all time scales; but in addition there are external influences like the changed luminosity of the sun, varying earth orbital parameters, impacts by comets, meteors and planetoids causing additional variations, which for most cases cannot be distinguished from internally caused variability as long as the history of the influencing factors is not well known. As illustrated in figure 1, the result is an impressive variability which even for the yearly mean hemispheric or global temperature reaches several tenths of a degree Celsius from one year to the next, often 0.4 °C in a decade or 0.8 °C for (taking 5 year averages) in the 20th century. If one plots precipitation of the rainy season, *the* fundamental parameter for the survival of populations in

* e-mail: grassl@dkrz.de

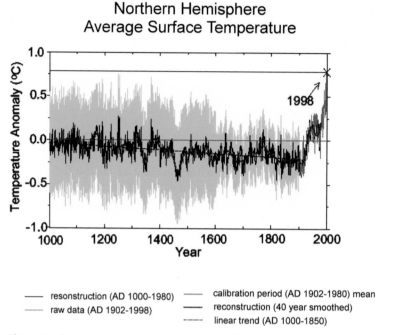

Fig. 1. Yearly mean 2 m air temperature of the Northern Hemisphere as derived from direct measurements and layered proxies [Source: Mann et al. (1999).]

many countries, year-to-year variability may reach more than 50 percent in the subhumid or semiarid tropics, indicating the enormous potential for development which would lie in the ability to give physically based probability estimates of the next rainy season, both for avoiding devastating impacts of severe droughts and for getting bumper crops in a good rainy season by adjusting governmental and agricultural management.

Understanding Interannual Variability

The major El Niño event 1982/83 stimulated co-ordinated research within the Tropical Ocean/Global Atmosphere (TOGA) project of WCRP, which started officially in 1985. The major goal was to understand the irregular oscillation in sea surface temperature in the tropical oceans but especially in the Eastern Pacific, which causes weather extremes and related natural disasters on nearly global scales; and to predict it if possible. The breakthrough to seasonal predictions for ENSO affected areas achieved by scientists from several nations working within TOGA can be seen as a major landmark for environmental sciences, since for the first time the probability for given climate anomalies on time scales far beyond the present 6 to 8 days for deterministic weather forecasting can be predicted by

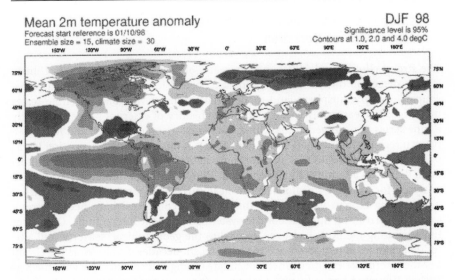

Fig. 2. The first global 3 months precipitation anomaly forecast issued by the European Centre for Medium-range Weather Forecast (ECMWF) for December, January and February 1997/98 during the recent major El Niño, based on observations gathered end of October 1998. Only areas were a 95 percent chance of a significant deviation from the mean was forecast are coloured.

supplying the proper starting field in atmosphere and upper ocean to coupled ocean/atmosphere models. These models simulate properly the intrinsic phenomena of the coupled tropical ocean/atmosphere system in the Pacific. To put it simply: Scientists have learned to transport and develop temperature anomalies in their models with the proper speed and partly the proper growth or decay rates, if an embryo of El Niño has been detected by the new subsurface observation system across the tropical Pacific. Since forecasts over several months and even over a few seasons are now available for many regions as probability estimates for a certain deviation from the mean from global 3-D coupled ocean/atmosphere models, mankind has entered a totally new era (see fig. 2). Much more sophisticated planning in many fields is now possible. Examples are: Adapted and often higher electricity production in major water reservoirs, changed agricultural practices even including changed crops, more adequate transport volumes for fossil fuels, higher disaster preparedness, changed behaviour on the stock market, improved holiday planning. If properly exploited, it might lead to faster development of areas suffering from strong natural climate variability. Seasonal forecasts can become an invaluable form of development aid.

Modelling Climate Variability as a Pre-requisite for Climate Change Detection

The strong climate variability on time-scales of years to decades makes climate change detection difficult. Four inputs are needed for detection: firstly a global coupled ocean/atmosphere/land-model, that simulates variability on time-scales up to decades reasonably well, secondly the history of external influence, thirdly a time series of observed global climate at least for some decades, and fourthly a pattern recognition method that is able to detect the emerging fingerprint. It was only in the mid-nineties that all these ingredients were available for the first thorough detection attempts (e.g. Hegerl et al. 1997). Although numerous time series analyses, for example for global mean near surface air temperature, had earlier pointed to a human influence on this record it could still have been a strong excursion driven by internal climate system component interactions. Only after fingerprinting methods had been applied to the geographical, seasonal and altitudinal changes of temperature during the recent decades did the Working Group I of the Intergovernmental Panel on Climate Change (IPCC) dealing with the Science of Climate Change formulate the now famous sentence: *The balance of evidence suggests a discernible human influence on global climate (IPCC 1996).* The main weaknesses still remaining were the inadequate history of radiative influence, for example for tropospheric sulphate aerosols, and model variability somewhat below the observed level for all models that jointly allowed for increased greenhouse effect and sulphate aerosols. The Second Assessment Report of IPCC thus had added a new pillar concerning anthropogenic climate change in addition to the earlier three: Observed concentration increase of long-lived greenhouse gases, model simulations of global warming for business as usual emission scenarios and past climate evidence for warming during periods with high greenhouse gas concentrations (for methane and carbon dioxide).

Attribution of Climate Change to Causes

After detection attribution is the next goal. As the increased greenhouse gas concentrations will cause cooling of the stratosphere, it seemed straight-forward to attribute the observed cooling there to this changed atmospheric composition.

However, the cooling trend during the recent decades was especially strong in the lower stratosphere for mid and high latitudes in altitudes where models would only point to the crossover from warming below to cooling above. Joint evaluations of ozone and temperature trends supported by model calculations have led to one of the first attributions of a climate change phenomenon to distinct causes: *The cooling in the lower stratosphere is largely due to ozone depletion and to a lesser extent to increased greenhouse gas concentrations*[1].

---

[1] WMO, UNEP (1999) assess ozone depletion regularly. The most recent assessment is published as Report No. 44 in the Global Ozone Research and Monitoring series of WMO, Geneva

WCRP, Interdisciplinarity and Transdisciplinarity

Part of the success of WCRP is due to the comparably low level of interdisciplinarity needed for some breakthroughs in the understanding of climate variability. Only two different disciplines of geosciences had to co-operate to get seasonal climate anomaly predictions for areas affected by the El Niño/Southern Oscillation phenomenon: meteorologists and physical oceanographers. The new tools they had to use jointly were: new observations of basic physical parameters at the ocean surface and in the upper ocean, partly from high-tech sensors, plus coupled ocean-atmosphere models of medium complexity. If the contribution of soil moisture to predictability on time-scales of weeks to months is sought another group of disciplines in geosciences have to co-operate, namely meteorology, hydrology and soil sciences.

The proper application of these predictions, however, still needs typical transdisciplinary research involving not only social sciences but also many categories of users. Therefore, I tried to encourage the START (Global Change *System* for *Analysis Research* and *Training*) project Climate Variability Predictions for Agricultural Production (CLIMAG), because it has all the ingredients needed to collate the parents of START (IHDP, IGBP, WCRP), to activate the regional START network to challenge the global projects of IGBP and WCRP, to bring in the farmers and the economists, the latter dealing with the food market over a larger area. A major challenge for WCRP projects caused by CLIMAG is: better regionalisation of further improved seasonal predictions. This means nesting of circulation models, ensemble predictions and dynamical down scaling in order to assess the probability of certain weather extremes, up to now hidden in an originally coarse seasonal prediction. On the other hand crop yield modellers (often just applying the model for a plot) have to upscale their results to a region and have to include the hitherto mostly omitted influence of pests and diseases into their models. The success of this transdisciplinary research in CLIMAG rests on excellent contributions in all disciplines, interdisciplinary co-operation, co-ordination across major groups of disciplines, acceptance by the users and joint funding by several nations, i.e., the stakes are very high. For the funding the major obstacle is the neglect of the correlation between standard of living and research/ development expenditure (see also section The Widening Gaps) in most developing countries.

The Inadequate Infrastructure of Some Global Change Research Programmes

Mankind has no chance to reach sustainable development in the coming century without global change research programmes, as clearly indicated by the detection of the Antarctic ozone hole and the phasing out chlorofluorocarbons and halones just in time to prevent a major disaster. But also the detection of global anthropogenic climate change (see IPCC 1966) has with high probability stimulated the acceptance of the Kyoto Protocol. This protocol of the United Nations Framework Convention on Climate Changes (UNFCCC) is the first major attempt to start global Earth system management by prescribing emission reductions for the

Table 1. Infrastructure for Global Change Research Programmes and an assessment of their adequacy

Programme	Date of Birth and Sponsors	Secretariat in	Number of Staff	Number of International project offices (IPOs)	Problems
World Climate Research Programme (WCRP)	1980 WMO, ICSU, IOC (since 1993)	Geneva at WMO	8	5	some IPOs too small
International Geosphere-Biosphere Programme (IGBP)	1986 ICSU	Stockholm Swedish Academy of Sciences	~10	9	some IPOs too small, basic funding endangered
DIVERSITAS Programme on Biodiversity Research	1991 UNESCO, several unions of ICSU	Paris at UNESCO	1	none	absolutely inadequate infrastructure
International Human Dimensions Programme of Global Environmental Change (IHDP)	1966 ICSU ISSC	Bonn at University	~5	1	inadequate infrastructure both for secretariat and IPOs

major consumers of fossil fuels after 200 years of continuous growth in fossil fuel use by the industrialised countries.

Are we prepared to detect further signals of global change that could threaten mankind early enough? If one looks at table 1 the answer is a disappointing "No" for some of the global environmental trends and their potential consequences. While WCRP can rely on an infrastructure which is in large parts just adequate with some problems in the area of international project offices (IPOs) which are not adequately funded (examples: GEWEX and SPARC), IGBP has recently suffered from a decline of funding by ICSU and has similar problems in the area of IPOs as WCRP. The real inadequacy of infrastructure becomes visible for the programmes DIVERSITAS and IHDP. Eight years after its foundation DIVERSITAS is still without any basic infrastructure needed for globally co-ordinated research projects. As this inadequacy is so obvious, the International Group of Funding Agencies (IGFA) supporting Global Change research programmes has decided in October 1999 in Beijing to help DIVERSITAS through a joint action. Hopefully deeds follow words. The youngest of the programmes (IHDP) is in the build-up phase which started when ICSU became sponsor besides the International Social Sciences Council (ISSC) in 1996, that had tried to attract the full scientific community into the HDP since 1992. The infrastructure is still not adequate and the secretariat in Bonn needs more international donors to guarantee continuing German support. Hopefully IPOs will be formed when the implementation plans follow the scientific plans for the present four projects.

The Widening Gap between the Major Research Nations and the Majority of Countries

When WCRP invited nations to the International CLIVAR Conference in December 1988 we expected that most nations would tell how they would contribute to the *Climate Variability* and Predictability study. However, about half of the 65 nations present asked WCRP to help develop means to translate improved climate variability predictions on seasonal to interannual time-scales to ease their use in a special area. In other words: potential donors declared themselves as recipients of anticipated research results.

The background for this behaviour is a widening gap between a few key research nations and the majority of nations not devoting enough resources to research and development. If developing countries will not increase the percentage of the gross national product devoted to research and development they have no chance to grow economically – in a sustained manner – faster than the upper rich part of the OECD countries. There is high correlation between living standard and research/development expenditure of a country. Even inside the European Union this relation holds. It hampers the reduction of differences in living standard if most of the Mediterranean EU members do not increase the research/development expenditure in relating to GNP to levels typical further North.

Why am I pointing to this basic problem? Because all our net-working in global change research, for example through the Global Change *System* for *Analysis, Research and Training* (START, jointly supported by IHDP, IGBP and WCRP) will not be successful in the long run if the developing nations will not shift expenditure from, for example, subsidies for fossil fuel to research and development. While the leading research nations give on average a bit less than 3 percent of GNP South America invests only 0.5 and Africa even only 0.2 percent; but also the Russian Federation fell to about 0.5 percent.

Earth System Analysis

Environmental Research already has acted as an early warning system for mankind. Examples are stratospheric ozone depletion, acidification of soils and freshwater systems, detection of anthropogenic climate change. In all the cases mentioned either the reversal of the trend or first internationally co-ordinated measures to dampen the growth rates have been reached. However, since we do not have a continuous global observing system besides the meteorological one, we do not know how many threats to the global environment and thus our development continue to be undetected. We need an *Earth System Analysis* based on global observations of physical, chemical, biological and socio-economic parameters, which then help to build improved models of the coupled system used for scenarios of possible futures. These scenarios are input for decision making. They have to be open for frequent revision, whenever more solid findings are available through better validation data sets for the models. Only in this manner can we pave the way to a sustainable future. Such an Earth System Analysis needs revised structures in research: Global programmes with sustained infrastructure, stronger co-ordination among the programmes, joint projects, strong interdisciplinarity, an open dialogue with decision makers of all groups in all regions, growing shares of GNP for education, research and development. Such an investment would be a low cost/high benefit undertaking. A shining example is for me the breakthrough to seasonal climate anomaly predictions. An investment of about 300 million US$ over 10 years is returned now to a single nation avoiding larger damage through preparedness for one El Niño event.

References

Hegerl G, Hasselmann K, Cubasch U, Mitchell JFB, Roeckner E, Voss R, Waskewitz J (1997) Multi-fingerprint detection and attribution analysis of greenhouse gas, gas-plus-aerosol and solar forced climate change. Climate Dynamics 13: 613–634

IPCC (1996) Second Full Assessment Report of the Intergovernmental Panel on Climate Change (IPCC) Working Group I Science of Climate Change. Cambridge University Press, Cambridge

CHAPTER 10

Earth System Models

Martin Claussen*

Earth system models: General remark

Earth system analysis – this term is often associated with the study of the 'solid' Earth with its surrounding spheres, the atmosphere, cryosphere, and hydrosphere. However, within IGBP (the International Geosphere – Biosphere Programme) – at least – a more general definition, which has been proposed by Schellnhuber (1998, 1999) and Claussen (1998), for example, seems to be generally accepted. According to the latter, Earth system analysis addresses the feedbacks and synergisms between the ecosphere and the anthroposphere. The ecosphere or, the natural Earth system, encompasses the abiotic world, the geosphere, and the living world, the biosphere, whereas the anthroposphere includes all cultural and socio-economic activities of humankind which can be subdivided into subcomponents such as the psycho-social sphere etc.

Schellnhuber (1998, 1999) provides a general **symbolic formalism of Earth system analysis** in the following way: The state of the natural Earth system is given by the vector N, which varies with time t. The evolution equation is F_0. The state of the anthroposphere, which is represented by some vector A, can be regarded as some time-dependent boundary conditions to the natural Earth system just as any other exogenous force E. Hence a model of the natural Earth system is schematically given by

$$\dot{N}(t) \equiv \frac{d\,N(t)}{dt} = F_0\left(N;\,t;\,(A(t),E(t))\right)$$

A similar equation for the state of the anthroposphere reads

$$\dot{A}(t) \equiv \frac{d\,A(t)}{dt} = G_0\,(N,\,A)$$

which could be considered a climate impact model, or, if G_0 would not depend on N, a scenario model, i.e., a model of population growth or increase of CO_2 emissions, for example.

Schellnhuber (1998, 1999, in this volulme) extends the above equations to set up a scheme of a fully coupled Earth system model. He uses this system not to aim at a forecast of the Earth's future, but to explore the system's behaviour in a

---

* e-mail: claussen@pik-potsdam.de

rather general way. For example, he designs a 'theatre world' to demonstrate the feasibility of measures and strategies of Earth system management.

How the equation, or system of equations, G_o, can be formulated explicitly, is not yet clear. The conventional, thermodynamic approach might not be the appropriate tool to model the anthroposphere. Presumably, anthropogenic activities are hardly assessable in term of thermodynamic quantities, i.e., they can hardly be cast into a mathematical form of state variables, state equations and evolution equations with the exception of, perhaps, economic aspects (e.g. Hasselmann et al. 1997, Klepper in this volume, for a detailed discussion of economic theories and coupling of climate and economic models). New methods of qualitative modelling have do be developed.

Even if these methods were to be developed, it is debatable whether the Earth system would be – even to a limited degree – predictable at all. Generally, people react to perceptions of global change rather than to the actual state of the Earth system. Hence any scientist, who tries to model the Earth system, readily interferes with the system by creating perceptions. Presumably, therefore, Earth system analysis will face the problem of a **global-scale uncertainty relation** (Zurek 1998). Just as at the micro scale, the interference of an observer with the system affects the system and, hence, the observation.

On the other hand, there are systems in physics and nature which are not predictable in principle, but manageable. For example, the dynamics of a coupled pendulum is not predictable, but reveals a deterministic chaos. Nonetheless, a child on a swing set – the 'daily-life realisation' of a chaotic coupled pendulum – is able to manage the system. Hence one of the major problems to be tackled by Earth system analysis is the proof that humankind can, or cannot, manage the system. This proof should not be considered as prerequisite for designing a new, better world. History suggests that, when trying to 'improve' our Earth, we most likely will face the same fate as Johann Wolfgang von Goethe's 'Sorcerer's Apprentice'. Instead, we ought to know whether we are able to curb the ongoing global-scale experiment or whether we have to adapt to our own 'mis-' management.

Modelling the Natural Earth System

Even if we ignore the interaction of the anthroposphere with the natural Earth system, modelling the natural Earth system still remains a challenge to which I will restrict myself in the following.

During the last 10,000 years, the present interglacial period, the climate has been rather calm in comparison with the previous 100,000 years (e.g. Dansgaard et al. 1993). Presumably, it is not a mere coincidence that agriculture has developed during the current phase of climatic stability. Generally the Earth's climate exhibits calm or slowly-varying conditions interspersed with episodes of rapid change on all time scales (e.g. Crowley and North 1991). Therefore, an important and exciting question is on how the conditions of little variability are maintained and what are the causes of rapid change. Moreover, we ought to know whether the natural Earth system could return to a more restless mode if anthropogenically induced modifications of the Earth continue to increase. It seems plausible

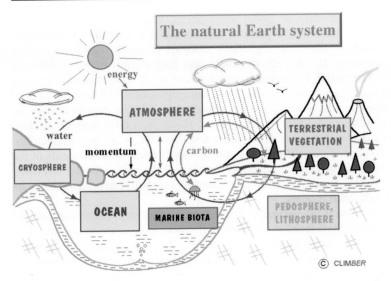

Fig. 1. Sketch of the natural Earth system. This sketch was designed by Andrey Ganopolski (Potsdam Institute for Climate Impact Research)

that changes in the North Atlantic thermohaline circulation could have caused drastic climate variability in the past (e.g. Bond et al. 1992) and that the same could happen in a warmer climate (Rahmstorf and Ganopolski 1999). Consequently, a major challenge for the scientific community today is to explore the dynamic behaviour of the natural Earth system as well as its resilience to large scale perturbation (such as the continuing release of fossil fuel combustion products into the atmosphere or the fragmentation of terrestrial vegetation cover).

To address the problem of stability of the natural Earth system one has to analyse the dynamic processes between its subsystems, the geosphere and the biosphere (e.g. Peixoto and Oort 1992). For this sake, the geosphere itself can be subdivided (see fig. 1) into the atmosphere, the hydrosphere (mainly the oceans), the cryosphere (inland ice, sea ice, and snow cover), the pedosphere (the soils), and the lithosphere (the Earth's crust and the more flexible upper Earth's mantle).

There is increasing evidence that the dynamics of the natural Earth system cannot be determined by studying its subsystems alone. Due to the (nonlinear) synergism between subsystems the response of the entire system to external perturbation drastically differs from the sum of the responses of the individual subsystem or a combination of a few of them – which will be demonstrated below.

Marked progress has been achieved during the past decades in modelling the separate elements of the geosphere and the biosphere (Houghton et al. 1997). This stimulated attempts to put all separate pieces together, first in form of **comprehensive coupled models** of atmospheric and oceanic circulation, and eventually as climate system models which include also biological and geochemical processes (Foley et al. 1998).

Comprehensive models of global atmospheric and oceanic circulation describe many details of the flow pattern, such as individual weather systems and regional currents in the ocean. Similarly, complex dynamic vegetation models explicitly determine the growth of plants and competition between different plant types. The major limitation in the application of comprehensive models arises from their high computational cost. The troposphere, the lowest 15 km of the atmosphere in which weather occurs, reacts within a few days to changes in boundary conditions, for example insolation. However, it takes several hundred years for the deep ocean to respond and a few thousand years to reach equilibrium. The response time will increase enormously if more 'slow' elements of the climate system, like glaciers or the upper Earth's mantle, are involved. Even using the most powerful computers, only a very limited number of experiments can be performed with such models.

Another problem is the necessity of ad hoc flux adjustments to obtain a realistic present climate state (see e.g. Cubasch et al. 1995). Flux adjustments are artificial corrections of simulated heat and freshwater fluxes at the interface between atmosphere and ocean models. The use of flux adjustments prevent the coupled atmosphere – ocean models from drifting into unrealistic climate states; however, they impose strong limitations on the applicability of the models to climate states which are substantially different from the present one.

At the other end of the spectrum of complexity of natural Earth-system models, we find **conceptual or tutorial models**. These models are simple mechanistic models which are designed to demonstrate the plausibility of processes. Watson and Lovelock's (1983) 'Daisyworld' model is just such an example. It provides a simple description of global-scale homoeostasis to show that biota could influence environment to form a self-regulating system in which conditions remain favourable for life. Watson and Lovelock (1983) do not claim that the biogeophysical process described in their conceptual model is realistic – it is just a caricature of the real world. Indeed, we have reasons to believe that the biogeophysical feedback, a positive temperature-albedo feedback, operating in the Daisyworld tends to destabilise the real-world natural Earth system. To cite a second example: Paillard (1998) describes the long-term climate variations during the last 1 million years, i.e., the variations between rather short interglacial and longer glacials, by assuming that there are multiple states in the natural Earth system. The system switches to the one or the other state, if changes in insolation exceed some *ad hoc* defined thresholds. Paillard does not make use of any physical constraints, he merely demonstrates in the most simple way the concept of thresholds in the natural Earth system.

Quite generally, conceptual models could be characterised as **inductive deterministic** models according to Saltzman (1985, 1988). They contrast with the comprehensive, **quasi-deductive** models which are, with respect to their main components, derived from first principles of hydrodynamics. Inductive deterministic models are formulated based on a gross understanding of the feedbacks that are likely to be involved. The system of equations – generally restricted to a very few – are designed to be capable of generating the known climatic variations, or as many lines of observational evidence as possible. The inductive approach is, to cite Saltzman (1985), *bound to be looked upon as nothing more that curve fit-*

ting – a charge that is fundamentally difficult to refute. Essentially, the predictive value of conceptual models is rather limited.

EMICs

To bridge the gap between conceptual and comprehensive models, **Earth System Models of Intermediate Complexity** (EMICs) have been proposed which can be characterise in the following way. EMICs describe most of the processes implicit in comprehensive models, albeit in a more reduced, i.e. a more parameterised form. They explicitly simulate the interactions among several components of the natural Earth system including biogeochemical cycles. On the other hand, EMICs are simple enough to allow for long-term climate simulations over several 10,000 years or even glacial cycles. Similar to comprehensive models, but in contrast to conceptual models, the degrees of freedom of an EMIC exceed the number of adjustable parameters by several orders of magnitude. EMICs are more quasi-deductive models, not inductive deterministic models, although some of the components of an EMIC could belong to this class. Indeed, most dynamic vegetation models are more or less empirically derived, inductive models. And also comprehensive models rely on inductive components to parameterise small scale processes.

Tentatively, we may define an EMIC in terms of a three-dimensional vector: Integration, i.e. number of components of the Earth system explicitly described in the model, number of processes explicitly described, and detail of description of processes (see fig. 2).

Currently, there are several EMICs in operation such as 2-dimensional, zonally averaged models (e.g. Gallée et al. 1991), 2.5-dimensional models with a simple energy balance (e.g. Marchal et. al 1998, Stocker et al. 1992), or with a statistical-

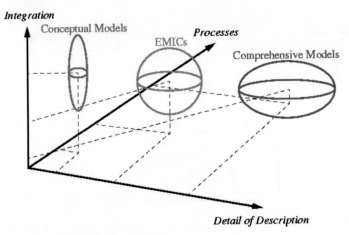

Fig. 2. Tentative definition of EMIC

dynamical atmospheric module (e.g. Petoukhov et al. 2000), and reduced-form comprehensive models (e.g. Opsteegh et al. 1998). Thus, there are a variety of EMICs of various degrees of complexity, and, obviously, there is no real gap between conceptual models, EMICs, and comprehensive models as suggested in figure 2.

The common denominator of all EMICs is their scope: all EMICs are designed as models describing the 'most important' processes governing the natural Earth system in order to facilitate long-term simulations or large-number ensemble simulations. A currently open question is which processes are 'most important' to properly simulate Earth system dynamics. Certainly, EMICs should more or less completely describe all components of the natural Earth system. Figure 3 depicts an example of the modular structure of an EMIC.

Moreover, all EMICs have to be global, geographically explicit models, because fluxes within the system are global (e.g., the hydrological cycle, the carbon cycle, and energy cycle) and changes in one region my well be caused by changes in a distant region. However, how much spatial resolution is required to appropriately capture processes with global significance?

In some cases, regional processes can be described in an aggregated form, as, for example, in so-called statistical-dynamical models of the atmosphere (Saltz-mann 1978, Petoukhov et al. 2000). Implicit in this approach is the assumption

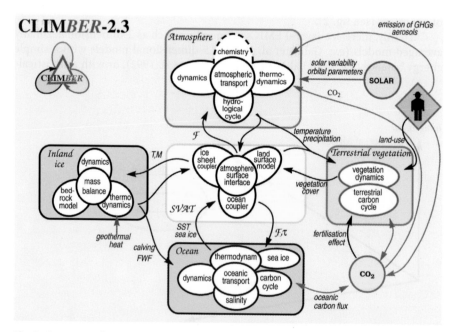

Fig. 3. Structure of CLIMBER-2.3, an EMIC developed at the Potsdam Institute for Climate Impact Research [Source: Petoukhov et al. (2000), Claussen et al. (1999).]

that the general structure of the atmosphere can be expressed in terms of large-scale, long-term fields of the main atmospheric variables, with characteristic spatial and temporal scales of $L > 1,000$ km and $T > 10$ days, and ensembles of synoptic-scale eddies and waves, i.e. weather systems like depressions, areas of high pressure, storms, etc., represented by their (L^2, T) averaged statistical characteristics. In other words, one parameterises the average transport effects of the rapidly varying weather systems on the large-scale, long-term atmospheric motion, rather than simulating them explicitly. This approach seems to be applicable to meridional heat and moisture transports from the subtropics to the high latitudes; however, it would lead to cumbersome hydrological pattern, if applied to low latitudes. There, moisture is advected from the arid subtropics towards the humid tropics – certainly not a diffusive process. In this case, an alternative model has to be found. For example, Petoukhov et al. (2000) prescribe the existence, but not amplitude and extent, of a Hadley cell regime, thereby allowing for counter-gradient meridional transports of moisture from the subtropics to the inter tropical convergence zone.

At the end, only extensive comparison with result from comprehensive models and validation against data and palaeo reconstructions will yield an answer on which processes are important to be included into an EMIC explicitly and which processes can be parameterised, i.e. described in an more aggregated, implicit way.

Examples using EMICs

EMICs have been used for a number of palaeostudies, because they provide the unique opportunity of transient, long-term ensemble simulations (e.g. Claussen et al. 1999b) – in contrast to so-called time slice simulations in which the climate system is implicitly assumed to be in equilibrium with external forcings – which rarely is a realistic assumption. Also the behaviour of the natural Earth system under various scenarios of greenhouse gas emissions has been investigated exploring the potential of abrupt changes in the system (e.g. Stocker and Schmittner 1997, Rahmstorf and Ganopolski 1999). Here is a brief summary of three examples of Earth system analysis using EMICs.

The next ice age

The current ice age, which the Earth entered 2 to 3 million years ago, is characterised by multiple switches of the global climate between glacials (with extensive ice sheets in the Northern Hemisphere) and interglacials (with climate similar to or warmer than today). The interglacials – at least during the last half million years of the so-called late Quaternary – were rather short in comparison with the glacials. The interglacial last some 10 ky (ky = 1,000 years), the glacials, some 100 ky. Our current interglacial, the Holocene, which peaked around 6 ky BP (before present), is already 11,5 ky old. Hence we face the question: when will we enter the next glacial?

Currently, there are two competing theories: Statistical extrapolation of palaeorecords (e.g. Thiede and Tiedemann 1998) suggest that the next ice age is just 'around the corner' and will start within the next thousand years. This argument is based on comparing the interglacials, in particular the last interglacial, the Eemian, which centred around 125 ky BP. Indeed, there is a striking similarity between the temperature curve of the Eemian and the Holocene. In particular, there seems to be a slight, but significant global long-term cooling trend from the mid-Holocene to today which could be interpreted as indication of the next ice age.

Following the astronomical theory of ice ages, however, the next glacial will not occur within the next 50 ky (e.g. Berger and Loutre 1997a, b). According to the astronomical theory, ice ages are triggered by changes in insolation at high northern latitudes, say at 65°N, during the summer solstice. These changes are caused by variations in the eccentricity of the Earth's orbit which has varied from near circularity to slight ellipticity at periods at about 100 ky and 400 ky over the last 3 million years. The tilt of the Earth's axis varied between 22° and 25° over a period of almost 41 ky. The wobble, i.e., the precession of the equinoxes, which is arises from the precession of the spinning Earth and the precession of the Earth's orbit, changes with a double period of 19 ky and 23 ky. In the past, large amplitude and high frequency variations in the summer insolation at 65° N have been observed. Starting some 60 ky BP, however, the precession cycle almost disappeared and between now and 50 ky AP (after present), the amplitude of variations in insolation is small owing to the very low value of eccentricity (Berger 1978, see fig. 4a).

The astronomical calculation suggests that the Eemian can not be an analogue of our present-day interglacial. Indeed, by using insolation changes as boundary conditions Berger and Loutre (1997a) predict that, in the absence of other forcings, such as anthropogenic greenhouse-gas emissions, the next glaciation in the northern hemisphere will occur about 55 ky AP (see fig. 4b). Hence Berger and Loutre (1997a) conclude that we are indeed going to the next ice age. However in contrast to the previous interglacials, the climate is likely to remain more or less stable and warm over the next 40 ky, then cools abruptly leading to the stadial of 55 ky AP, and finally to the next glacial maximum of 100 ky AP.

The current warm phase: the Holocene

Remnants of the last glaciation had disappeared by about 7000 years ago and since then, the inland ice masses have changed little. Nevertheless, the climate was quite different from today's climate. Generally, the summer in Northern Hemisphere mid- to high latitudes was warmer as palaeobotanic data indicate an expansion of boreal forests north of the modern treeline (Foley et al. 1994). In North Africa, palaeoclimatological reconstructions using ancient lake sediments and archaeological evidence indicate a climate wetter than today (Yu and Harrison 1996). Moreover, it has been found from fossil pollen that the vegetation limit between Sahara and Sahel reached at least as far north as 23°N (Jolly et al. 1998).

It is hypothesised that differences between modern and mid-Holocene climate were caused by changes in the Earth's orbit (Kutzbach and Guetter 1986). Particularly, the tilt of the Earth's axis was stronger than today. This led to an increased solar radiation in the Northern Hemisphere during summer which amplified the

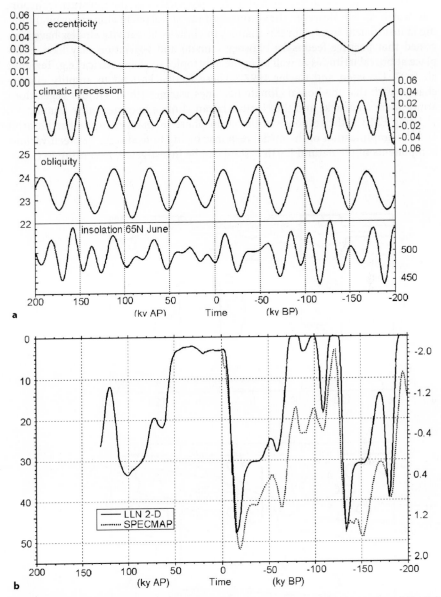

Fig. 4. a Long-term variations of the astronomical elements (eccentricity, precession, obliquity) and of the insolation at 65°N at the summer solstice from 200 ky BP to 200 ky AP. [Source: Berger (1978) with modifications.] **b** Full, thick line: Long-term variations of the northern hemisphere ice volume (in 10^6 km³) simulated by Berger and Loutre (1997a). (Please note that ice volume increases downward, left hand scale.) Dashed, thin line: changes in the oxygen isotope ^{18}O, a proxy for ice volume change on global average. [Source: Berger and Loutre (1997a) with modifications.]

African and Indian summer monsoon, thereby increasing the moisture transport into North Africa. However, the response of the atmosphere alone to orbital forcing is insufficient to explain the changes in climate. Sensitivity studies have suggested that positive feedbacks between climate and vegetation may have taken place at boreal latitudes as well as in the subtropics of North Africa (e.g. Texier et al. 1997, Claussen and Gayler 1997). These feedbacks tend to amplify climate change such that the boreal climate becomes warmer (than without vegetation-atmosphere feedback) and the North African climate becomes more humid.

Using EMICs, the response of the atmosphere to changes in the Earth's orbit and the amplification of the initial response owing to feedbacks between various components of the natural Earth system can be investigated. According to Gano-

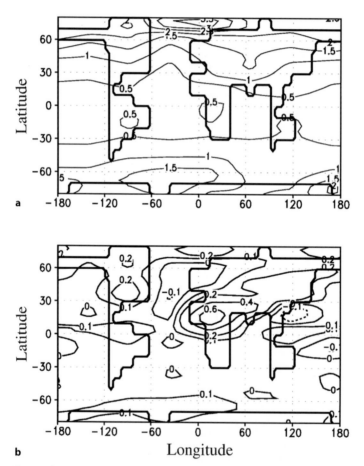

Fig. 5. Changes in annual air temperature (in ° C) near the Earth's surface (**a**) and annual precipitation (in mm/day) (**b**) during the mid-Holocene, 6000 years before present, compared to today. [Source: Claussen et al. (1999b).]

polski et al. (1998), it appears that the atmospheric response to orbital forcing alone yields a summer warming and a winter cooling above the Northern Hemisphere continents. If terrestrial vegetation interacts with the atmosphere, then a much stronger warming is found over the northern continents. This can be explained by a northward shift of forests and associated so-called taiga-tundra feedback. The biogeophysical feedback reduces the albedo in spring and early summer as snow-covered forests appear to be much darker than snow-covered grassland, thereby absorbing more solar radiation. If the atmosphere-vegetation system is coupled with the ocean, then a further temperature increase in summer and a warming instead of a cooling in winter is observed. On annual average, the warming over the Northern Hemisphere reaches up to 4°C (see fig. 5a). The additional warming is caused by a stronger reduction of Arctic sea ice owing to the synergism between taiga-tundra feedback and sea-ice-albedo feedback which often, but not quite correctly is referred to as the 'biome paradox'. Precipitation differences are strongest over North Africa (see fig. 5b), mainly owing to the atmosphere-vegetation interaction.

During the last several thousand years, the climate has changed to a cooler and more arid state in which the present-day subtropical deserts fully developed. How did this long-term climate change happen, was it gradual or did it occur in steps? By using an EMIC, one finds that changes in high northern latitudes, i.e., changes in the abundance of taiga and tundra, occurred rather gradual. However, in the subtropics of North Africa, climate developed more abrupt – in comparison with the external forcing, the change in the Earth's orbit around the sun. According to Claussen et al. (1999a), climate and vegetation reacted smoothly to the external forcing until some 5,500 years ago. Then it changed rapidly, followed by a further gradual drift (see fig. 6).

A rather rapid decrease of Saharan vegetation, even more rapid than indicated in figure 6, has recently been reconstructed from aeolian dust transport from the Western Sahara into the subtropical North Atlantic (de Menocal et al. 2000). In the more continental position of the Eastern Sahara, the desertification was pre-

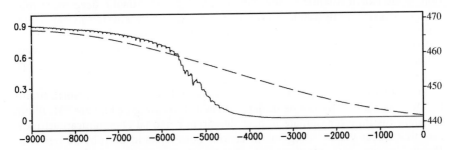

Fig. 6. Development of vegetation fraction in the Sahara (full line, non-dimensional units, left ordinate) as response to changes in insolation of the northern hemisphere during boreal summer (dashed line, in W/m², right ordinate). The abscissa indicates the number of years before present. [Source: Claussen (2000).]

sumably not as fast as in the western part, where the North African wet phase ended around 5,000 to 4,500 years (e.g. Pachur and Wünnemann 1996).

This study of the natural Earth system dynamics in the (geologically) recent past suggests that Saharan desertification at the end of the mid-Holocene was presumably a natural phenomenon. Deforestation by neolithic people apparently occurred in some places (Pachur, personal communication), but Saharan desertification can be explained without taking it. On the other side, the strong climate and vegetation change should have had a profound impact on the neolithic society. It has been suggested that the foundation of high civilisation along the Nile river was influence, perhaps even dominated by people's migration from the increasingly arid Sahara to the more fertile banks of river Nile.

The threat of abrupt climate change

Often, anthropogenically induced climate change is viewed upon as steady global warming. However, future climate change may also come as unexpected, large and rapid climate system changes.

Palaeoclimatic evidence suggests (e.g. Bond et al. 1993) that some past climate shifts were associated with changes in the formation of North Atlantic deep water. There is a consensus today that the thermohaline circulation of the World Ocean is to a large degree driven from the high-latitude North Atlantic through the production of North Atlantic deep water. Sinking of surface water in the Greenland-Iceland-Norwegian Sea and in the Labrador Sea initiates an overturning-circulation cell on the meridional plane in which northward transport of upper-ocean warm water is balanced by deep return flow of cold water, imposing a strong northward heat flux in both the North and South Atlantic. This heating system makes the northern North Atlantic about 4 °C warmer than corresponding latitudes in the Pacific and is responsible for the mild climate of Western Europe. Variations in North Atlantic deep water formation therefore have the potential to cause significant climate change in the North Atlantic region.

Moreover, the pioneering work by Stommel (1961) suggests that the thermohaline circulation is a non-linear system which is highly sensitive to freshwater forcing such as changes in meltwater flow from glaciers or changes in precipitation in the North Atlantic region. Formation of North Atlantic deep water may collapse if a certain threshold is exceeded and it can show a hysteresis behaviour. There are studies (e.g. Manabe and Stouffer 1994, Stocker and Schmittner 1997, Rahmstorf and Ganopolski 1999) which indicate that abrupt changes in North Atlantic deep water formation could happen if anthropogenic emissions of greenhouse gases continue to rise unchecked (see fig. 7).

The potential impacts of an abrupt decrease in the production of North Atlantic Deep Water can hardly be underestimated. First rough estimates (M. Blum, Potsdam Institute for Climate Impact Research, personal communication) indicate that wheat yields in regions of central Europe could drop by 50 per cent. Hence it seems desirable to strictly avoid such drastic changes in the natural Earth system. The work by Stocker and Schmittner (1997) provide some guidance, how this can be achieved. Their work suggests that the onset of an abrupt decrease in North Atlantic Deep Water formation depends not only the amplitude

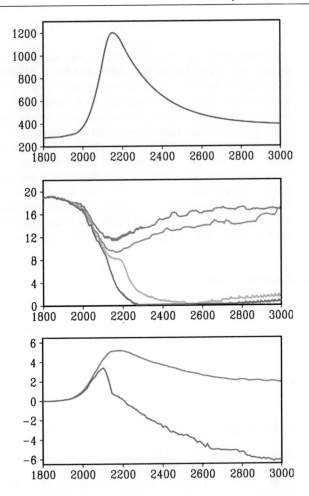

Fig. 7. Upper figure: The CO_2-forcing starting with observed CO_2-concentration, continuing according to the IPCC IS92e scenario to the year 2100. After 2100, CO_2-concentrations are assumed to peak in 2150 and to decline thereafter. Middle figure: The response of the simulated thermohaline circulation without any artificial freshwater input into the North Atlantic (green line). The blue, orange, and red curves are scenarios with additional freshwater input (+0.1 Sv, +0.15 Sv, +0.2 Sv, respectively.1 Sv = 1 Sverdrup = 10^6 m³/s). Lower figure: Change in winter temperatures over the North Atlantic and North-West Europe region (right figure). Source: Rahmstorf and Ganopolski (1999) with modifications.

of CO_2-emissions, but also on the rate of emissions. The Earth system stays on the 'safe' side, i.e., with North Atlantic Deep Water forming, when the rate of emissions is sufficiently slow – which is an important conclusion with respect to global Earth system management.

Summary

Analysis of the natural Earth system generally relies on a hierarchy of simulation models. Depending on the nature of questions asked and the pertinent time scales, there are, on the one extreme, zero-dimensional tutorial or conceptual models like those in the 'Daisyworld family'. At the other extreme, three-dimensional comprehensive models, e.g. coupled atmospheric and oceanic circulation with explicit geography and high spatio-temporal resolution, are under development in several groups.

During the IGBP (International Geosphere-Biosphere Programme) Congress in Shonan Village, Japan, May 1999, and the IGBP workshop on EMICs in Potsdam, Germany, June 1999, it became more widely recognised that models of intermediate complexity could be very valuable in exploring the interactions between all components of the natural Earth system, and that the results could be a more realistic than those from conceptual models. These meetings have pointed at the potential that EMICs even might have for the policy guidance process, such as the IPCC (the Intergovernmental Panel on Climate Change).

Finally, it should be emphasised that EMICs are considered to be one part of the above mentioned hierarchy of simulation models. EMICs are not likely to replace comprehensive nor conceptual models, but they offer a unique possibility to investigate interactions and feedbacks at the large scale while largely maintaining the geographic integrity of the natural Earth system.

Acknowledgements

I would like to thank the CLIMBER group, Andrey Ganopolski, Vladimir Petoukhov, Victor Brovkin, Claudia Kubatzki, Stefan Rahmstorf, Reinhard Calov, Eva Bauer, Anja Hünerbein, and Irina Fast, for great team work. Furthermore, I very much appreciate constructive co-operation and discussion with John Schellnhuber and Wolfgang Cramer, PIK, and André Berger, Université Catholique, Louvain la Neuve, Belgium.

References

Berger A (1978) Long-term variations of daily insolation and Quaternary climatic changes. J Atmos Sci 35: 2362–2367
Berger A, Loutre MF (1997a) Palaeoclimate sensitivity to CO_2 and insolation. Ambio 26 (1): 32–37
Berger A, Loutre MF (1997b) Long-term variations in insolation and their effects on climate, the LLN experiments. Surveys in Geophysics 18: 147–161
Bond G, Broecker W, Johnsen, McManus J, Labeyrie, Jouzel J, Bonani G (1993) Correlations between climate records from North Atlantic sediments and Greenland ice. Nature 365: 143–147
Claussen M (1998) Von der Klimamodellierung zur Erdsystemmodellierung: Konzepte und erste Versuche. Annalen der Meteorologie (NF) 36: 119–130
Claussen M (2000) Biogeophysical feedbacks and the dynamics of climate. In: Schulze ED, Harrison SP, Heimann M, Holland EA, Lloyd J, Prentice IC, Schiimel D (eds) Global Biogeochemical Cycles in the Climate System. Academic Press, San Diego, in press
Claussen M, Gayler V (1997) The greening of Sahara during the mid-Holocene: results of an interactive atmosphere – biome model. Global Ecology and Biogeography Letters 6: 369–377
Claussen M, Kubatzki C, Brovkin V, Ganopolski A, Hoelzmann P, Pachur HJ (1999a) Simulation of an abrupt change in Saharan vegetation at the end of the mid-Holocene. Geophys Res Letters 24 (14): 2037–2040
Claussen M, Brovkin V, Ganopolski A, Kubatzki C, Petoukov V, Rahmstorf S (1999b) A new model for climate system analysis. Env Mod Assmt 4: 209–216

Crowley T, North G (1991) Paleoclimatology. Oxford Monographs on Geology and Geophysics 18, Oxford University Press, New York

Cubasch U, Santer BD, Hegerl GC (1995) Klimamodelle – wo stehen wir? Phys Bl 51: 269–276

Dansgaard W, Johnsen SJ, Clausen HB, Dahl-Jensen D, Gundestrup NS, Hammer CU, Hvidberg CS, Steffensen JP, Sveinbjörnsdottir AE, Jouzel J, Bond G (1993) Evidence for general instability of past climate from a 250-kyr ice-core record. Nature 364: 218–220

Foley J, Kutzbach JE, Coe MT, Levis S (1994) Feedbacks between climate and boreal forests during the Holocene epoch. Nature 371: 52–54

Foley JA, Levis S, Prentice IC, Pollard D, Thompson SL (1998) Coupling dynamic models of climate and vegetation. Global Change Biology 4: 561–580

Gallée H, van Ypersele JP, Fichefet T, Tricot C, Berger A (1991) Simulation of the last glacial cycle by a coupled, sectorially averaged climate–ice sheet model. I: The climate model. J Geophys Res 96 (13): 139–161

Ganopolski A, Kubatzki C, Claussen M, Brovkin V, Petoukhov V (1998) The Influence of vegetation-atmosphere-ocean interaction on climate during the mid-Holocene. Science 280: 1916–1919

Hasselmann K, Hasselmann S, Giering R, Ocana V, von Storch H (1997) Sensitivity study of optimal CO_2 emission paths using a simplified structural integrated assessment model (SIAM). Climatic Change 37: 345–386

Houghton JT, Meira Filho LG, Griggs DJ, Maskell K (1997) An introduction to simple climate models used in the IPCC second assessment report. IPCC Technical Paper II

Jolly D, Harrison SP, Damnati B, Bonnefille R (1998) Simulated climate and biomes of Africa during the late quarternary: comparison with pollen and lake status data. Quaternary Science Reviews 17 (6–7): 629–657

Kutzbach JE, Guetter PJ (1986) The influence of changing orbital parameters and surface boundary conditions on climate simulations for the past 18,000 years. J Atmos Sci 43: 1726–1759

Manabe S, Stouffer R (1994) Multiple-century response of a coupled ocean-atmosphere model to an increase of atmospheric carbon dioxide. J Clim 7: 5–23

Marchal O, Stocker TF, Joos F (1998) A latitude-depth, circulation-biogeochemical ocean model for paleoclimate studies: model development and sensitivities. Tellus 50B: 290–316

de Menocal PB, Ortiz J, Guilderson T, Adkins J, Sarnthein M, Baker L, Yarusinski, M (2000) Abrupt onset and termination of the African Humid Period: Rapid climate response to gradual insolation forcing. Quat Sci Rev 19: 347–361

Opsteegh JD, Haarsma RJ, Selten FM, Kattenberg A (1998) ECBILT: A dynamic alternative to mixed boundary conditions in ocean models. Tellus 50A: 348–367

Pachur HJ, Wünnemann B (1996) Reconstruction of the palaeoclimate along 300E in the eastern Sahara during the Pleistocene/Holocene transition. In: Heine K (ed) Palaeoecology of Africa and the surrounding islands, pp 1–32

Paillard D (1998) The timing of Pleistocene glaciations from a simple multiple-state climate model. Nature 391: 378–38

Peixoto JP, Oort AH (1992) Physics of Climate. American Institute of Physics, New York

Petoukhov V, Ganopolski A, Brovkin V, Claussen M, Eliseev A, Kubatzki C, Rahmstorf S (2000) CLIMBER-2: a climate system model of intermediate complexity. Part I: Model description and performance for present climate. Climate Dyn 16 (1): 1–17

Rahmstorf S, Ganopolski A (1999) Long-term global warming scenarios computed with an efficient coupled climate model. Climatic Change 43: 353–367

Saltzman B (1978) A survey of statistical-dynamical models of the terrestrial climate. Adv Geophys 20: 183–304

Saltzman B (1985) Paleoclimatic modeling. In: Hecht AD (ed) Paleoclimate analysis and modeling. John Wiley & Sons Inc., pp 341–396

Saltzmann B (1988) Modelling the slow climatic attractor. In: Schlesinger ME (ed) Physically-based modelling and simulation of climate and climatic change. Part II. Kluwer Academic Publishers, pp 737–754

Schellnhuber HJ (1998) Discourse: Earth System Analysis – The Scope of the Challenge. In: Schellnhuber HJ, Wenzel V (eds) Earth System Analysis – Integrating science for sustainability. Springer, Heidelberg, pp 5–195

Schellnhuber HJ (1999) 'Earth system' analysis and the second Copernican revolution. Nature 402: C19-C23

Stocker TF, Schmittner A (1997) Influence of CO_2 emission rates on the stability of the thermohaline circulation. Nature 388: 862–865

Stocker TF, Wright DG, Mysak LA (1992) A zonally averaged, coupled ocean-atmosphere model for paleoclimate studies. J Climate 5: 773–797

Stommel H (1961) Thermohaline Convection with two stable regimes of flow. Tellus 13: 225–230

Texier D, de Noblet N, Harrison SP, Haxeltine A, Jolly D, Joussaume S, Laarif F, Prentice IC, Tarasov P (1997) Quantifying the role of biosphere-atmosphere feedbacks in climate change: coupled model simulations for 6000 years BP and comparison with palaeodata for northern Eurasia and northern Africa. Climate Dyn 13: 865–882

Thiede J, Tiedemann R (1998) Die Alternative: Natürliche Klimaveränderungen – Umkippen zu einer neuen Kaltzeit? In: Lozán JL, Graßl H, Hupfer P (eds) Warnsignal Klima/Wissenschaftliche Fakten. Wissenschaftliche Auswertungen, Hamburg, S 190–196

Watson AJ, Lovelock JE (1983) Biological Homeostasis of the Global Environment: The Parable of Daisyworld. Tellus 35B: 284–289

Yu G, Harrison SP (1996) An evaluation of the simulated water balance of Eurasia and northern Africa at 6000 y BP using lake status data. Climate Dynamics 12: 723–735

Zurek WH (1998) Decoherence, Chaos, Quantum-Classical Correspondence, and the Algorithmic Arrow of Time. Physica Scripta T76: 186

CHAPTER 11

Climate System and Carbon Cycle Feedback   **11**

Pierre Friedlingstein*

CO$_2$ in the climate system

Ice core data show a strong correlation between atmospheric CO$_2$ and global temperature over the glacial cycles, indicating that the climate system is closely coupled to the carbon cycle. During the four last glacial cycles, over the past 420,000 years, atmospheric CO$_2$ had excursions from 200 ppmv during cold glacial periods, up to 280 ppmv during warm interglacial periods (Petit et al. 1999) (see fig. 1a). Other trace gases such as methane also show a strong temporal coupling with temperature over glacial cycles (Petit et al. 1999). Over the more recent history, atmospheric CO$_2$ has been recorded to increase from roughly 280 ppmv at the dawn of the industrial revolution, up the 360 ppmv today (see fig. 1b) (Etheridge et al. 1996). This sharp increase is due to a) the burning of fossil fuel for energy production (Andres et al. 1996) and b) the intense deforestation, essentially in the tropics, needed to meet the increasing food and fibre demand (Houghton 1995). Many forests have been cut or degraded, and today, a large fraction of the terrestrial ecosystems is directly influenced by human activities. Land use over the past 200 years has caused terrestrial ecosystems to release carbon (mainly to the atmosphere).

Within the same time, the global mean surface temperature has been increasing by about 0.7°C, 1997 and 1998 successively breaking the records of warmest years (Karl et al. 2000). Whether this global warming is due to the increase of atmospheric CO$_2$ is still a subject for hot debate, although, quoting the Intergovernmental Panel on Climate Change (IPCC) climate change 1995 report, "The body of statistical evidence, when examined in the context of our physical understanding of the climate system, now points towards a discernible human influence on global climate" (Santer et al. 1996a). The most advanced coupled atmosphere-ocean general circulation models (AOGCM) are used to simulate the historical evolution of the climate and its link to human activity (Santer et al. 1996b, Bengtsson et al. 1999) but also the future climate change that would occur if the rate of release of greenhouse gases, such as carbon dioxide, methane, nitrous oxide, etc still increase in the future (e.g. Kattenberg et al. 1996) (see fig. 2). Although a large degree of uncertainty remains in these climate predictions, several patterns are robust (Houghton et al. 1996). For a given greenhouse gases

---

* e-mail: pierre@lsce.saclay.cea.fr

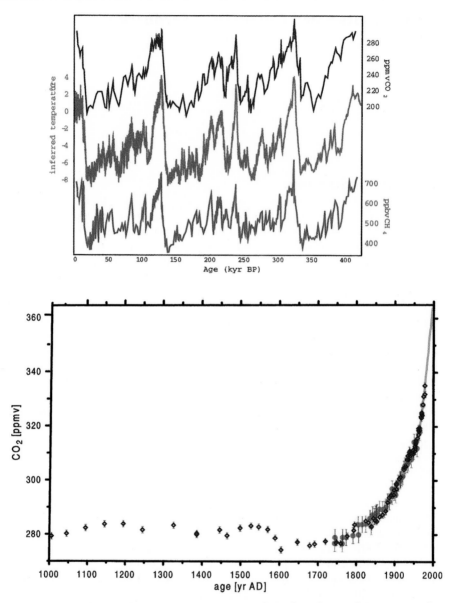

Fig. 1 a. Time evolution of isotopic temperature, CO_2 and CH_4 from the Vostok ice core over the last 400 10^3 years. [Source: Petit et al. (1999).] **b.** Combined ice core and atmospheric CO_2 measurements showing the last deglaciation (from 20 to 10 10^3 years before present) and the rapid increase over the last century (http://www.pages.unibe.ch/publications/overheads.html).

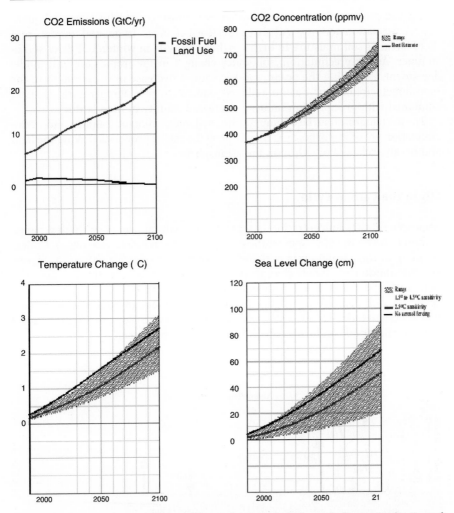

Fig. 2. Intergovernmental Panel on Climate Change calculations of climate implication of anthropogenic CO_2 emissions. Time series of a) IS92a scenario of CO_2 emission, b) calculated atmospheric CO_2 concentration for the IS92a emissions, c) calculated temperature change with respect to present-day, and d) calculated sea level change. Purple zone represents the range of estimation. [Source: Houghton et al. (1996). (http://ipcc-ddc.cru.uea.ac.uk/cru-data/examine/emissions/is92a–graphs.html).]

emission scenario (e.g. IPCC IS92 scenarios), models predict a global warming ranging from 1 °C to 3.5 °C by the end of the 21st century. Warming on lands is generally larger than on the oceans, the largest being in high northern latitudes. All models produce an increase in global mean precipitation, although the spatial pattern, especially in the tropics, changes from model to model. As a result from changes in precipitation and evaporation, soil moisture, which is a crucial quantity for vegetation activity, will also experience changes. In general, models predict increased soil moisture in high northern latitude in winter, but drier soil in summer. As for precipitation, soil water changes in the tropics are still model dependent. Regarding the ocean, models show a decrease in the strength of the meridional circulation, due to reduced equator-to-pole temperature gradient combined with precipitation induced decreased salinity in high latitudes.

These changes of surface temperature, soil water content, salinity and ocean circulation are of importance as they control the major processes driving the land and ocean exchange of CO_2 with the atmosphere.

CO_2 in the Carbon Cycle

Atmospheric CO_2 concentration is regulated by natural processes that exchange carbon between the atmosphere, the ocean and the land biota (see fig. 3). Understanding of the carbon cycle is therefore necessary to predict future CO_2 levels and the consequent climate change. In the following, I will present first the "natural" carbon cycle as it stood during pre-industrial times, and then analyse the "perturbed" carbon cycle, as it is presently driven out of equilibrium by sustained human induced CO_2 emissions. This will help us to comprehend the future of the carbon cycle and its feedback with the climate system.

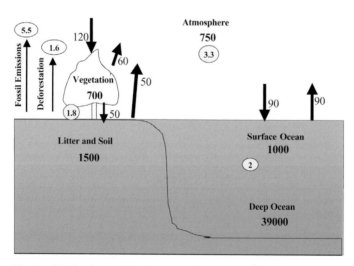

Fig. 3. Global carbon cycle. Fluxes are in GtC/yr (10^{12} kgC/yr), pools are in GtC. Circled numbers represent the anthropogenic perturbation fluxes (fossil fuel and deforestation CO_2 emissions), the accumulation in the atmosphere, and the uptakes by the land and the ocean. Source: adapted from Houghton et al. (1996).

The pre-industrial carbon cycle

The ocean is by far the largest carbon pool with 40,000 GtC (compared with 750 GtC stored in the atmosphere and 2,200 GtC on lands). CO_2 is exchanged with the atmosphere because of air-sea CO_2 partial pressure difference. As atmospheric CO_2 concentration is quite homogeneous, air-sea exchanges are primarily function of surface water CO_2 partial pressure (pCO_2) and gas exchange velocity. CO_2 in surface water is controlled by temperature (CO_2 solubility decreases with temperature), by oceanic circulation (through the carbonate equilibrium), but also by biological activity. The ocean is physically stratified, so that only carbon dissolved into the surface layer is exchanged with the atmosphere. The mixing between surface waters and the deep sea is much slower, requiring decades to centuries. The rate of vertical mixing is important as the deep ocean is clearly the main long term reservoir for anthropogenic CO_2. Future changes of oceanic circulation are therefore extremely important for the carbon cycle and the estimate of the future CO_2 airborne fraction.

On lands, plants and soils contain carbon both in living biomass (leaves, branches, stems and roots) and in soils (litter, soil organic matter). The overall size of the terrestrial biosphere is about 2,200 GtC. Carbon is distributed among various pools, each associated with a residence time that ranges between a few months (leaves on deciduous trees) to few hundred years (organic matter in peat). Atmospheric CO_2 is fixed through photosynthesis, which depends mainly on air temperature, water stress and atmospheric CO_2. This flux, the gross primary productivity (GPP) amounts to 120 GtC/yr. A large fraction (50 %) of GPP is respired back to the atmosphere by the living plants through autothrophic respiration. The remaining fraction carbon, known as net primary productivity (NPP) is stored by the plants as tissues in the different compartments (leaf, stems, roots). The turnover time of carbon in these tissues vary from a few months (leaves of annual plants and deciduous trees) to several decades (stems of boreal coniferous trees). Dead biomass accumulates as litter and soil organic matter, these pools being respired by decomposer organisms. The amount of carbon stored above and below ground is function of the ecosystem type (e.g. grassland vs. forest) but also function of climate. Indeed tropical regions, where warm and wet conditions prevail, show high rate of organic matter decomposition, as opposed to boreal regions where decomposer activity is strongly inhibited during the winter period. As a result, high latitude ecosystems, such as tundra have the largest below-ground carbon stock although the annual accumulation (through NPP) is much lower than in the tropics.

As for the ocean, this concept of residence time of carbon in the system is of importance as it will directly translate into potential storage of anthropogenic carbon. An ecosystem with a very fast carbon turnover time is unlikely to be an important carbon sink, as the in and out fluxes will always be close to equilibrium. Therefore, any climate induced change in carbon turnover time (e.g. increased organic matter decomposition rate due to higher soil temperature) will impact on the terrestrial carbon cycle and on atmospheric CO_2.

The industrial carbon cycle

Since human beings started burning fossil fuel and harvesting forests, atmospheric CO_2 has steadily increased. However, is has been shown that the cumulated emissions since the pre-industrial time (1850) is about twice as large as the measured increased of CO_2 in the atmosphere (e.g. Houghton et al. 1996). That is to say, there is a negative feedback occurring in the Earth System: the increase of atmospheric CO_2 enhances the net CO_2 flux from the atmosphere to the land and/ or to the ocean, and therefore limits the rate of atmospheric increase. The mechanisms responsible for the uptake of excess CO_2, the partitioning between land and ocean uptake, and more important, the regional distribution of this carbon sink are still poorly understood (Fan et al. 1998, Bousquet et al. 1999). Table 1 and figure 3 give the best estimate of the different compounds of the present-day carbon budget (Houghton et al. 1996).

The atmospheric storage is very accurately measured worldwide at more than 50 different sites (Conway et al. 1994). Fossil emissions of CO_2 are derived from fossil fuel production statistics for each country with accuracy on the order of 10 % in industrialised countries (Andres et al. 1996). The ocean storage is estimated by different independent ways: a) global carbon cycle models (Orr 1999), b) extrapolation of air-sea pCO_2 measurements (Takahashi et al. 1997), c) extrapolation of ocean carbon inventory (Gruber et al. 1996), or d) atmospheric measurements of several tracers (CO_2, O_2/N_2) (Keeling et al. 1996, Battle et al. 2000). The land net storage is deduced as the difference between fossil fuel source and the sum of atmosphere and ocean storage. As shown in table 1, this term is close to zero, i.e., the biosphere is almost neutral (at least for the 1980's). However this "neutral" biosphere actually hides two opposite terms: the deforestation induced CO_2 flux, that amounts 1.6 ±1. GtC/yr, and the land uptake that can only be derived by difference. This term is difficult, if not impossible, to measure directly, the terrestrial biosphere being extremely heterogeneous, the net CO_2 fluxes are variable in space and time. Direct measurements, using eddy-correlation techniques, allow to quantify net ecosystem CO_2 fluxes over a small area (1–5 km) (e.g. Goulden et al. 1996, Baldocchi et al. 1996), however the extrapolation to large continental areas is not realistic yet. The land uptake of carbon can only be estimated using (1) global models of the terrestrial biogeochemical carbon cycle,

Table 1. Estimate of the global budget for the 1980's and for the 1990's. The fluxes are given in GtC/yr.

	1980 to 1989	1989 to 1998
(1) Fossil emissions	5.5 ± 0.5	6.3 ± 0.6
(2) Atmospheric storage	3.3 ± 0.2	3.3 ± 0.2
(3) Ocean storage	2.0 ± 0.8	2.3 ± 0.8
(4) Land net storage: (1)-[(2)+(3)]	0.2 ± 1.0	0.7 ± 1.0
(5) Deforestation source	1.6 ± 1.	1.2 ± 1.
(6) Derived land uptake= (4)-(5)	1.8 ± 1.4	1.9 ± 1.4

Source: After IPCC (1996) and update.

and (2) indirect inferences from atmospheric tracers such as CO_2, O_2/N_2 and CO_2 isotopes. Several factors are expected to control carbon storage by land ecosystems: the CO_2 fertilisation effect on photosynthesis, the nitrogen deposition impact on productivity, the effects of climate change and variability on carbon fluxes, and the changes in disturbance regimes. Again, none of these processes has been directly measured at the global scale, we have to rely on small scale measurements (from leaf level to site level experiments) and on terrestrial model estimates based on "our best understanding" of these processes. Although these processes are inter-dependent and may very well combine in a non-linear way, it has been attempted to quantify their respective contribution to the global land CO_2 uptake (Schimel 1995). It is crucial to bear in mind that any attempt to model future climate system and carbon cycle will directly rely on the assumptions adopted to represent the carbon cycle and especially the land uptake component. A necessary condition for such prediction is obviously to be able to produce a realistic carbon budget for the historical period.

Climate-Biogeochemistry Feedbacks

On land, biospheric activity, which is mainly driven by the climate, exerts a strong control on the fluxes of water, energy and CO_2. Hence, there are numerous feedbacks between the land biogeochemistry and the atmosphere. The same does not apply to the ocean where biological activity has little control on sea-air fluxes of water and energy. The only parallel between land and ocean applies for the CO_2 fluxes, which in both case are controlled by biogeochemistry, driven by the climate. The land-atmosphere feedbacks operate at different scales, from the short-term local scale (e.g. diurnal coupling between the vegetation and the boundary layer) to the continental scale (e.g. Amazon deforestation) and the global scale (e.g. carbon cycle). A very extended review and analysis of local scale feedbacks can be found in Pielke et al. (1998) and in Raupach (1998). In the following, I will concentrate only on large scale mechanisms. Figure 4 attempts to represent the main feedbacks occurring between the land ecosystems and the atmosphere. Although they may occur simultaneously, they can be decomposed into 4 separate pathways.

Physiological feedback

Increased atmospheric CO_2 is known to affect plant physiology. Indeed, plants have "openings" on their leaves, called stomates, which connect the plant to the atmosphere for the exchange of water and CO_2. CO_2 increase will induce a stomatal closure, that is to say, stomatal conductance decreases with CO_2 (e.g. Field et al. 1995). This mechanism can be explained from an "economical" view: stomatal conductance controls the plant in-flux of CO_2 for photosynthesis, but also the out-flux of water from transpiration. If ambient CO_2 increases, closing of the stomate will drastically reduce the water loss without affecting the CO_2 gain when compared to today's CO_2 condition, i.e. the plant water use efficiency increases with CO_2 (Drake et al. 1997). Hence the feedback: increased CO_2, through

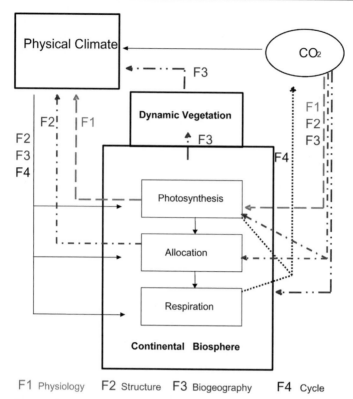

F1 Physiology   F2 Structure   F3 Biogeography   F4 Cycle

Fig. 4. Schematic view of the four main feedbacks between the terrestrial biosphere and the climate system. The green, red, pink and blue arrows describe the physiological, structural, biogeographical and carbon cycle feedbacks respectively, as described in the text.

reduced stomatal conductance, induces a reduction of plant transpiration. Earlier studies quantified the climate response to a given change, typically a doubling, in stomatal resistance (Henderson-Sellers et al. 1993a, Pollard and Thompson 1995). In 1996, Sellers et al. put a step forward and simulated a doubling of CO_2 with an atmospheric GCM coupled to a land surface scheme that includes a representation of photosynthesis (Sellers et al. 1996). Doing so, they were in a position to evaluate the direct effect of atmospheric CO_2 on climate via the standard radiative effect and the indirect effect of CO_2 on climate through calculated changes in stomatal conductance together with the impact of climate and CO_2 on plant assimilation of carbon. They showed that at $2xCO_2$, decreased in stomatal conductance induces a decrease of transpiration and an increase in air temperature amplifying the changes due to the standard radiative effect. In the tropics, the physiological feedback leads to a 0.4 °C extra warming on the 1.7 °C warming due to the radiative effect. GPP on the other hand is mainly boosted by elevated CO_2 (+32 %) while the warming alone only increase tropical GPP by 1 %.

Structural feedback

Shortly after Sellers, Betts et al. (1997) confirmed the physiological feedback but highlighted a second mechanism that comes into the game, playing in the opposite direction than Sellers' mechanism. As CO_2 increases, plant productivity is boosted, leading to larger stand biomass. Of importance is leaf biomass (generally expressed as leaf area index, LAI, the leaf surface normalised by the ground surface). LAI increase will enhance transpiration, as the exchange surface is larger. So, elevated CO_2 reduces stomatal conductance, but enhances stomates amount, hence the two opposite effects on transpiration. In their study, Betts et al. found that the two effects tend to cancel out, their net effect on surface temperature being lower than what Sellers found earlier. Another potentially important structural feedback that received little attention so far is the change in roots distribution. Climate change and increase in CO_2 have the potential to affect allocation to roots (van Noordwijk et al. 1998, Friedlingstein et al. 1999), which in turn impacts on climate through changes in evapotranspiration (Kleidon and Heimann 1998, Kleidon et al. 2000). In a model experiment, Kleidon and Heimann (1998) calculate the rooting depth that is optimum for plant productivity and estimate its control on the hydrological cycle and the climate system. Dynamic roots being much larger than what is used conventionally in GCMs, the transpiration rate is larger and sensible heat flux and surface temperature are reduced. Structural changes in plant partitioning may play an important role in future climate-land interaction.

Biogeographical feedback

Increased CO_2 and climate change may affect ecosystem functioning to the limit where a given ecosystem is not competitive anymore. That means ecosystem dieback, migration, colonisation. All of these climate induced land cover changes will result in feedback on the atmosphere through changes in albedo, water and energy fluxes. Several studies showed the strong link between land cover distribution and climate. One of the earliest one was Charney's study (1975) on desert which feedbacks upon itself, through high albedo, high radiative heat loss, strong descending air branches, and low relative humidity. More recent studies focussing on specific ecosystems highlighted the importance of the biogeographical feedbacks. In the boreal region, it has been showed that replacement of tundra by boreal forest, as happened in the mid-Holocene, through a reduction in surface albedo, contributed to the climate change at that time (Foley et al. 1994). In short, the boreal forest contributes to the relatively warm conditions of the boreal regions (Bonan et al. 1992, Chalita and LeTreut 1994). Several studies also focused on tropical forests where large deforestation occurs, and is believed to affect the climate not only through release of CO_2, but also directly through changes in water and energy exchanges with the atmosphere (e.g. Henderson-Sellers et al. 1993b, Polcher and Laval 1994). The most recent study combines deforestation with increased atmospheric CO_2 to quantify, using a atmospheric GCM coupled to a biogeochemical land surface model, their respective impact on the future climate of Amazon basin (Costa and Foley 2000). The authors found that both deforestation and elevated CO_2 tend to increase surface temperature, but they have opposite effects on precipitation. Other simulations, coupling dynamic vegetation models with GCMs investigated the consistency of climate and vegetation distribution (e.g. Claussen 1998, Foley et al. 1998) or the impact of the change from potential to actual vegetation on climate (Chase et al. 1996, Bonan 1997, Stohlgren et al. 1998). Finally a recent study, although only conceptual, showed as an extreme sensitivity study how the climate would be different if the surface were uniformly green (evergreen forest) or covered by deserts (Kleidon et al. 2000).

Carbon cycle feedback

Climate change and elevated CO_2 affect photosynthesis, allocation, plant and soil respiration, ... As a result, the net CO_2 flux between the land and the atmosphere may depart from equilibrium, inducing a change in atmospheric CO_2 concentration that will feedback on the climate system. As mentioned in section 2.2, the impact of elevated CO_2 on photosynthesis is believed to already act in today's carbon cycle and to be partially responsible for the land uptake of CO_2 (Houghton et al. 1996). The same applies for the ocean, elevated CO_2 enhances diffusive air to sea flux. These atmospheric CO_2-carbon cycle negative feedbacks have been estimated for the present or for the future by several models (e.g. Kicklighter et al. 1999, Friedlingstein et al. 1995, Orr 1999).

Equally important is the climate change impact on the carbon cycle. Several studies showed that future climate may tend to affect the biospheric uptake. One of the earliest study was performed by Melillo et al. (1993) where a terrestrial model was forced by both future CO_2 and future climate. Since that time, many terrestrial models have been run under climate change conditions at regional scales (VEMAP members 1995, White et al. 2000) or at the global scale (King et al. 1997, Cao and Woodward 1998). Climate change is generally found to cause reduction in net biospheric uptake. The reason for this negative impact has to be found in the tropical regions where higher temperature and increased drought reduce plant productivity but increase heterotrophic respiration. In an extreme case, this eventually leads to a severe dieback of tropical forest which further increase carbon loss from the biosphere (Cramer et al. 2000). Temperate and high latitude regions seems, on the contrary do benefit from the climate change (Cao and Woodward 1998, White et al. 2000).

Separate ocean studies have similar results, climate change impact on ocean circulation and sea surface temperature will tend to reduce the global ocean carbon uptake (Maier-Reimer et al. 1996, Sarmiento et al. 1996, 1998, Joos et al. 1999, Matear and Hirst 1999).

These land and ocean mechanisms, if realistic, represent a strong positive feedback in the Earth system: elevated CO_2 induces a climate change that inhibit land and ocean carbon uptakes, leaving a larger fraction of CO_2 in the atmosphere, and therefore accelerating the climate change. Ongoing studies undertaken at the Institute Pierre Simon Laplace (IPSL, Paris) and at the United Kingdom Meteorological Office (U.K. Met. Office) are coupling GCMs and global carbon cycle models. Preliminary results are indeed confirming this positive feedback.

Combining the feedbacks

The vision of four independent feedbacks, as described above is, of course, theoretical. In the real world they may occur simultaneously, or at least gradually, the
system being in essence non linear, its global response will be anything but the
sum of the individual responses. Mooney and Koch (1994) and Field et al. (1995)
sketched the interactions between CO_2, atmosphere dynamic, and ecosystem
functioning. Based on their precursory works, figure 5 attempts to combine these
CO_2-climate-ecosystems feedback loops. In spite of its apparent complexity, figure 5 is still an oversimplified vision of the real system. For example, feedbacks
involving non-CO_2 greenhouse gases (CH_4, N_2O) or atmospheric chemical compounds (NOx, O_3) are still absent of this picture.

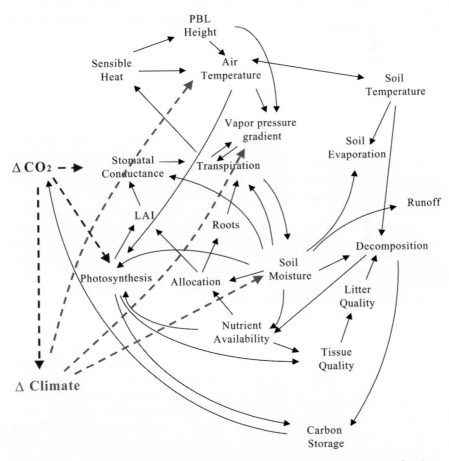

Fig. 5. Interactions between the climate system, the surface physic and the biogeochemistry.
[Source: adapted from Mooney and Koch (1994), and Field et al. (1995).]

Future Directions

Predictions of future climate, as shortly described in section 1, require a set of calculations:
(a) the development of atmospheric CO_2 emission scenarios,
(b) a conversion of CO_2 emissions into atmospheric CO_2 concentration, and
(c) the calculation of climate change for the given change in atmospheric CO_2,

the same applying for all other greenhouse gases. Different methods can be adopted to achieve this goal. The usual approach, which has been extensively used (e.g. Houghton et al. 1996) is to perform the calculations of (a) emissions, (b) concentration, and (c) climate change one by one (see fig. 2). This has the advantage of simplicity but has a severe drawback: any feedback between the bio-geochemical cycles and the physical climate system, as described above, are not addressed in such approach as each calculation is done independently. Among the inconsistencies, the conversion of CO_2 emissions into CO_2 concentration is performed assuming the future carbon cycle operates under a present-day climate, although the ultimate goal is to estimate a climate change. Along the same lines, future CO_2 emission scenarios include a deforestation term, but the GCM estimates the climate change assuming a present-day vegetation distribution.

A more realistic approach, adopted by a few groups (IPSL and U.K. Met. Office), attempts to simulate climate and carbon cycle change simultaneously by the use of climate models coupled to land and ocean carbon cycle models. The forcing variable of these climate and carbon cycle models (CCCM) is therefore CO_2 emissions, conversion from emission to concentration being calculated online by the coupled model, accounting for the changing climate and its potential impact on the carbon cycle. The CCCM approach is obviously more realistic than the standard approach, used by IPCC, however, it is much more computer time consuming, and it requires a high degree of interaction from a large group of scientists from different fields (atmospheric physics, ocean physics, ocean bio-geochemistry, plant physiology).

However, as mentioned before, the CCCM approach should only be seen as a first step toward Earth system modelling. Other important trace gases are still ignored in this approach such as CH_4, NMHC, N_2O, DMS, aerosols, etc. All these species have natural sources controlled by climate, and play a role in the radiative budget of the Earth. There is still a long way before integrating these compounds in a common model framework.

Understanding of the Earth system and its response to the anthropogenic perturbation is a complex problem. As showed in this paper, there are several feedbacks operating at different time and spatial scales that have to be quantified in order to get a more realistic representation of the present and future climate system. The quantification of these feedbacks, their upscaling from the leaf or stand level to the GCM grid, and integration in global models are crucial research actions for future climate predictions.

Acknowledgements

I wish to thank L. Bengtsson who introduced me to the Nationales Komitee für Global Change Forschung (NKGCF) conference "Understanding the Earth System: Compartments, Processes and Interactions". I also want to thank M. Berthelot, L. Bopp, P. Ciais, N. deNoblet, L. Fairhead, J.-L. Dufresne, H. LeTreut, and P. Monfray for our continuous interaction on this subject.

References

Andres RJ, Marland G, Fung I, Matthews E (1996) A one degree by one degree distribution of carbon dioxide emissions from fossil-fuel consumption and cement manufacture, 1950–1990. Glob Biogeochem Cycles 10: 419–429

Baldocchi D, Valentini R, Running S, Oechel W, Dahlman R (1996) Strategies for measuring and modelling carbon dioxide and water vapour fluxes over terrestrial ecosystems. Global Change Biol 3: 159–168

Battle M, Bender M, Tans PP, White JWC, Ellis JT, Conway T, Francey RJ (2000) Global carbon sinks and their variability inferred from atmospheric O_2 and $\delta^{13}C$. Science 287: 2467–2470

Bengtsson L, Roeckner E, Stendel M (1999) Why is global warming proceeding much slower than expected? J Geophys Res 104: 3865–3876

Betts RA, Cox PM, Lee SE, Woodward FI (1997) Contrasting physiological and structural vegetation feedbacks in climate change simulations. Nature 387: 796–799

Bonan GB (1997) Effects of land use on the climate of the United States. Climatic Change 37: 449–486

Bonan GB, Pollard D, Thompson SL (1992) Effect of boreal forest vegetation on global climate. Nature 359: 716–718

Bousquet P, Ciais P, Peylin P, Monfray P (1999) Optimisation of annual atmospheric CO_2 net sources and sinks using inverse modelling. Part 1: method and control inversion. J Geophys. Res 104: 26161–26178

Cao M, Woodward FI (1998) Dynamic responses of terrestrial ecosystem carbon cycling to global climate change. Nature 393: 249–252

Chalita S, LeTreut H (1994) The albedo of temperate and boreal forest and the Northern Hemisphere climate: a sensitivity experiment using the LMD GCM. Climate Dynamics 10: 231–240

Charney JG (1975) Dynamics of deserts and drought in the Sahel. Quart J R Met Soc 101: 193–202

Chase TN, Pielke RA, Kittel TGF, Nemani R, Running SW (1996) Sensitivity of a general circulation model to global changes in leaf area index. J Geophys Let 101: 7393–7408

Claussen M (1998) On multiple solutions of the atmosphere-vegetation system in present-day climate. Glob Change Biolog 4: 549–560

Conway TJ, Tans PP, Waterman LS, Thoning KW, Kitzis DR, Masarie KA, Zhang N (1994) Evidence for interannual variability of the carbon cycle from the National Oceanic and Atmospheric Administration/Climate Monitoring and Diagnostic Laboratory Global Air Sampling Network. J Geophys Res 99: 22831–22855

Costa MH, Foley JA (2000) Combined effects of deforestation and doubled atmospheric CO_2 concentrations on the climate of Amazonia. J Climate 13: 18–34

Cramer W, Blondeau A, Woodward FI, Prentice IC, Betts RA, Brovkin VB, Cox PM, Fisher V, Foley JA, Friend AD, Kucharik C, Lomas MR, Ramankutty N, Sitch S, Smith B, White A, Young-Molling C (2000) Global response of terrestrial ecosystem structure and function to CO_2 and climate change: results from six dynamic global vegetation models. Glob Change Biolog, in press

Drake BG, Gonzalez-Meler MA, Long SP (1997) More efficient plants: A consequence of rising atmospheric CO_2? Annu Rev Plant Physiol Mol Biol 48: 609–639

Etheridge DM, Steele LP, Fangenfelds RL, Francey RJ, Barnola J-M, Morgan VI (1996) Natural and anthropogenic changes in atmospheric CO_2 over the last 1000 years from air in Antarctic ice and firn. J Geophys Res 101: 4115–4128

Fan S, Gloor M, Mahlman J, Pacala S, Sarmiento J, Takahashi T, Tans P (1998) Atmospheric and oceanic CO_2 data and models imply a large terrestrial carbon sink in North America. Science 282: 442–446

Field CB, Jackson RB, Mooney HA (1995) Stomatal responses to increased CO_2: implications from the plant to the global scale. Plant Cell and Environm 18: 1214–1225

Foley JA, Kutzbach JE, Coe MT, Levis S (1994) Feedbacks between climate and boreal forests during the Holocene epoch. Nature 371: 52–54

Foley JA, Levis S, Prentice IC, Pollard D, Thompson SL (1998) Coupling dynamic models of climate and vegetation. Glob Change Biolog 4: 561–580

Friedlingstein P, Fung I, Holland E, John J, Brasseur G, Erickson D, Schimel D (1995) On the contribution of CO_2 fertilization to the missing biospheric sink. Glob Biogeochem Cycles 9: 541–556

Friedlingstein P, Joel G, Field CB, Fung IY (1999) Toward an allocation scheme for global terrestrial carbon models. Glob Change Biolog 5: 755–770

Goulden ML, Munger JW, Fan SM, Daube BC, Wofsy SC (1996) Exchange of carbon dioxide by a deciduous forest: Response to interannual climate variability. Science 271: 1576–1578

Gruber N, Sarmiento JL, Stocker TF (1996) An improved method for detecting anthropogenic CO_2 in the oceans. Glob Biogeochem Cycles 10: 809–837

Henderson-Sellers A, McGuffie K, Gross C (1993a) Sensitivity of global model simulations to increased stomatal conductance and CO_2 increases. J Climate 8: 1738–1756

Henderson-Sellers A, Dickinson RE, Durbridge TB, Kennedy PJ, McGuffie K, Pitman AJ (1993b) Tropical deforestation: Modeling local- and regional-scale climate change. J Geophys Res 98: 7289–7315

Houghton JT, Meira Filho LG, Callander BA, Harris N, Kattenberg A, Maskell K. (eds) (1996) Climate Change 1995. The Science of Climate Change, Contribution of Working Group I to the Second Assessment Report of the Intergovernmental Panel on Climate Change. Cambridge University Press, Cambridge

Houghton RA (1995) Land-use change and the carbon cycle. Glob Change Biolog 1: 275–287

Joos F, Plattner GK, Stocker TF, Marchal O, Schmittner A (1999) Global warming and marine carbon cycle feedbacks on future atmospheric CO_2. Science 284: 464–467

Karl TR, Knight RW, Baker B (2000) The record breaking global temperatures of 1997 and 1998: Evidence for an increase in the rate of global warming? Geophys Res Let 27: 719–722

Kattenberg A, Giorgi F, Grassl H, Meehl GA, Mitchell JFB, Stouffer RJ, Tokioka T, Weaver AJ, Wigley TML (1996) Climate models – Projections of future climate. In: Houghton JT, Meira Filho LG, Callander BA, Harris N, Kattenberg A, Maskell K. (eds) Climate Change 1995. The Science of Climate Change, Contribution of Working Group I to the Second Assessment Report of the Intergovernmental Panel on Climate Change. Cambridge University Press, Cambridge, pp. 285–357

Keeling RF, Piper SC, Heimann M (1996) Global and hemispheric CO_2 sinks deduced from changes in atmospheric O_2 concentration. Nature 381: 218–221

Kicklighter DW, Bruno M, Donges S, Esser G, Heimann M, Helfrich J, Ift F, Joos F, Kaduk J, Kohlmaier GH, McGuire AD, Melillo JM, Meyer R, Moore B, Nadler A, Prentice IC, Sauf W, Schloss A, Sitch S, Wittenberg U, Wurth G (1999) A first-order analysis of the potential role of CO_2 fertilization to affect the global carbon budget: a comparison of four terrestrial biosphere models. Tellus 51B: 346–366

King AW, Post WM, Wullschleger S (1997) The potential response of terrestrial carbon storage to changes in climate and atmospheric CO_2. Climatic Change 35: 199–227

Kleidon A, Heimann M (1998) Optimised rooting depth and its impacts on the simulated climate of an Atmospheric General Circulation Model. Geophys Res Let 25: 345–348

Kleidon A, Fraedrich K, Heimann M (2000) A green planet versus a desert world: estimating the maximum effect of vegetation on the land surface climate. Climate Change 44: 471–493

Maier-Reimer E, Mikolajewicz U, Winguth A (1996) Future ocean uptake of CO_2: interaction between ocean circulation and biology. Clim Dyn 12: 711–721

Matear RJ, Hirst AC (1999) Climate change feedback on the future oceanic CO_2 uptake. Tellus 51B: 722–733

Melillo JM, McGuire AD, Kicklighter DW, Moore B, Vorosmarty CJ, Schloss AL (1993) Global climate change and terrestrial net primary production. Nature 363: 234–240

Mooney HA, Koch GW (1994) The impact of rising CO_2 concentrations on the terrestrial biosphere. Ambio 23: 74–76

Orr JC (1999) Ocean Carbon-Cycle Model Intercomparison Project (OCMIP): Phase 1 (1995–1997), IGBP/GAIM Report 7

Petit JR, Jouzel J, Raynaud D, Barkov NI, Barnola J-M, Basile I, Bender M, Chappellaz J, Davis M, Delaygue G, Delmotte M, Kotlyakov VM, Legrand M, Lipenkov VY, Lorius C, Pepin L, Ritz C, Saltzman E, Stievenard M (1999) Climate and atmospheric history of the past 420,000 years from the Vostok ice core, Antarctica. Nature 399: 429–436

Pielke RA, Avissar R, Raupach M, Dolman AJ, Zeng X, Denning S (1998) Interactions between the atmosphere and terrestrial ecosystems: influence on weather and climate. Glob Change Biolog 4: 461–475

Polcher J, Laval K (1994) The impact of African and Amazonian deforestation on tropical climate. J Hydrol 155: 389–405

Pollard D, Thompson SL (1995) Use of a land surface transfer scheme (LSX) in a global climate model: the response to doubling stomatal conductance. Glob Planetar Change 10: 129–161

Raupach MR (1998) Influences of local feedbacks on land-air exchanges of energy and carbon. Glob Change Biolog 4: 477–494

Santer BD, Wigley TML, Barnett TP, Anyamba E, and Contributors (1996a) Detection of climate change and attribution of causes. In: Houghton JT, Meira Filho LG, Callander BA, Harris N, Kattenberg A, Maskell K. (eds) Climate Change 1995. The Science of Climate Change, Contribution of Working Group I to the Second Assessment Report of the Intergovernmental Panel on Climate Change. Cambridge University Press, Cambridge, pp. 407–443

Santer BD, Taylor KE, Wigley TML, Johns TC, Jones PD, Karoly DJ, Mitchell JFB, Oort AH, Penner JE, Ramaswamy V, Schwarzkopf MD, Stouffer RJ, Tett S (1996b) A search for human influences on the thermal structure of the atmosphere. Nature 382: 39–46

Sarmiento JL, Le Quéré C (1996) Oceanic carbon dioxide uptake in a model of century-scale global warming. Science 274: 1346–1350

Sarmiento JL, Hughes TMC, Stouffer RJ, Manabe S (1998) Simulated response of the ocean carbon cycle to anthropogenic climate warming. Nature 393: 245–249

Schimel DS (1995) Terrestrial ecosystems and the carbon cycle. Glob Change Biol 1: 77–91

Sellers PJ, Bounoua L, Collatz GJ, Randall DA, Dazlich DA, Los SO, Berry JA, Fung I, Tucker CJ, Field CB, Jensen TG (1996) Comparison of radiative and physiological effects of doubled atmospheric CO_2 on climate. Science 271: 1402–1406

Stohlgren TJ, Chase TN, Pielke RA, Kittel TG, Baron JS (1998) Evidence that local land use practices influence regional climate, vegetation and stream flow patterns in adjacent naturel areas. Glob Change Biolog 4: 495–504

Takahashi T, Feely RA, Weiss RF, Wanninkhof RH, Chipman DW, Sutherland ST, Takahashi TT (1997) Global air-sea flux of CO_2: an estimate based on measurements of sea-air pCO_2 difference. Proc Natl Acad Sci 94: 8292–8299

van Noordwijk M, Martikainen P, Bottner P, Cuevas E, Rouland C, Dhillion SS (1998) Global change and root function. Glob Change Biolog 4: 759–772

VEMAP Members (1995) Vegetation/ecosystem modeling and analysis project: Comparing biogeography and biogeochemistry models in a continental-scale study of terrestrial ecosystem responses to climate change and CO_2 doubling. Glob Biogeochem Cycles 9: 407–437

White A, Cannel MGR, Friend AD (2000) The high-latitude terrestrial carbon sink: a model analysis. Glob Change Biolog 6: 227–245

CHAPTER 12

Global Climate Change and Economic Analysis 12

GERNOT KLEPPER*

Although very volatile, the climate system of the earth has been varying within broad but clearly defined bounds (see the chapter by Claussen in this book). It was not until the 18th century when economic activities started to change this. By today the atmospheric concentrations of greenhouse gases, and among them carbon dioxide (CO_2), methane (CH_4) and nitrous oxide (N_2O) have surpassed preindustrial levels. The CO_2-concentration in the atmosphere has grown from 280 to about 360 ppbv and is approaching 400 ppbv which is far beyond the concentrations that have been observed in the last half million years. CH_4 has grown from 700 to 1720 ppbv and N_2O from 275 to 310 ppbv (IPCC 1995).

Climate variability and climate change have therefore changed in the last one hundred years as the industrial revolution has spread throughout the world. The rising production has been brought about by technological development, most notably by the advances made in using new and alternative energy sources. Figure 1 illustrates the composition of the fuel supply since the mid 19th century.

The corners indicate the three sources oil/gas, coal, and renewables such as wood and water power plus nuclear energy. The figure 1 illustrates nicely the 19th century switch from the renewable energy source wood to coal and since the 1920s the subsequent introduction of oil and later gas which has continued until the 1970s with over 50 % of all energy consumption coming from oil and gas. In the last 20 years there has been little change in the composition of energy sources. The question, of course, is in which direction future energy supplies will go since this will determine future CO_2-emissions. The figure also illustrates one possible future energy path which we will come back to later.

The economic system has thus in a continuously changing way made its impact on the global as well as regional climate system and will continue to do so in the future. However, climate models indicate that a continuation of the current trend in emissions will most likely lead to drastic and unprecedentedly fast changes in the climate system to which ecosystems will not be able to adapt in time and which might impose economic problems at least to the more vulnerable regions of the earth. There is little disagreement that these current trends should and can not continue. However, for a large number of reasons there is no agree-

---

* e-mail: gklepper@ifw.uni-kiel.de

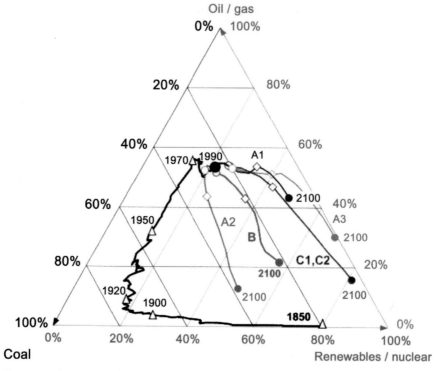

Fig. 1. Development of the Energy Mix. [Source: Nakicenovic et al. (1998).]

ment as to how and when the economic system should change course. One of the major reasons is the fact that economics has been faced with a number of new research questions which are now in the process of being resolved.

The Economic System and the Climate System

Economic activities influence in a complex way the climate system. The current pattern of energy consumption and of energy technologies used is the result of a historic process which has emerged as the combined outcome of an increasing shortage of renewable energy sources, namely wood, of new technological developments which allowed coal and later oil and gas to become major fossil energy sources. This cheap and large energy supply in turn has accelerated economic activity, productivity, and thus higher standards of living. These have contributed to higher life-expectancy and increased population growth which in turn has increased energy demand. Despite remarkable improvements in energy productivity these positive feedback effects have lead to a rising use of fossil energy sources, to increasing emissions of greenhouse gases, to high CO_2-concentrations

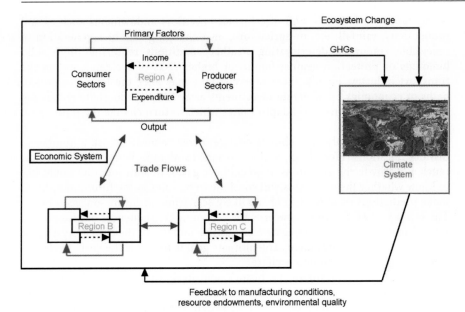

Fig. 2. The Economic and the Climate System

in the atmosphere, but so far not to a climate change that lies strongly beyond the natural variability of the climate system.

The analysis of the interaction of the economic system with the climate system in its most simple form is illustrated in figure 2. Economic activities in the different regions of the world lead to emissions of greenhouse gases and to changes in the ecosystems through land use changes or local pollution. The climate system reacts by changing climate conditions which may have an impact on the manufacturing conditions, on the availability of resources such as land or renewable resources, but also on the amenities provided by the nature system to humankind.

This interaction between the economic and the climate system has long time-lags. Whereas the reaction of the climate system to emissions has a delay of more than half a century, economic development and economic planning have time horizons which rarely reach beyond a few decades. As a consequence a set of new research question has been raised in economics and other disciplines which are related to the determinants of long-run economic development.

Research Questions

The central role of economic activities in the interplay between the natural and societal systems illustrated in figure 2 is visible most clearly in the first order effect of the emissions of greenhouse gases which set in motion the whole chain of climate change effects. The second order effects are composed of the repercussions which climate change imposes on the economic system through a deteriorating supply of crucial – and often sector-specific – endowments such as soil productivity, rainfall, etc. in agriculture, or a changing rate of regeneration of renewable resources, deteriorating weather conditions requiring more solid buildings or protection against floods, or higher risks of accidents, e.g. in the transport industry. Finally, these repercussions may alter the long-run growth paths of economics by diverting investments away from productivity enhancing ones towards protective and adaptive ones which in turn change the future path of emissions.

The economic analysis of climate change therefore needs to ask a number of distinct questions. First of all, it is important to identify who contributes how much and why to the emissions of greenhouse gases. In addition, one would like to know whether these insights will continue to be valid in the future in order to make predictions or at least to create likely scenarios for future emission paths. The second set of questions concerns the economic consequences of historic and future emission in terms of negative or positive impacts of the climate system on the human living environment. The ultimate goal is to assess the costs of climate change in terms of economic welfare.

Once an understanding of the historic and likely future development of the economic system, its impact on the climate system as well as the repercussions from the climate system on the economic system has been improved, the normative question needs to be raised: What should be done about climate change, provided it is seen as an undesirable situation? A desirable time profile of emissions of greenhouse gases should be determined which maximizes future welfare appropriately defined. Such an emission profile needs to be implemented through economic instruments and it needs to find a more or less unanimous support internationally since greenhouse gases are a global phenomenon which can only be addressed effectively through an internationally coordinated approach. Since different policy instruments imply a differing distribution of the costs of emission reductions, an analysis of the welfare effects of alternative instruments on a regionally disaggregated level is desirable for identifying politically acceptable instruments.

Modeling Approaches

It goes without saying that the grand design of a complex integrated model addressing all the above mentioned research questions simultaneously is impossible to achieve. Different approaches on different levels of abstraction and with different methods need to be used for addressing the above mentioned questions. Economic analysis and economic modeling can be of two types: It can describe the actual interplay of agents where each one maximizes her individual objective function given the constraints imposed on them by technology, institutions and rules, but also by the natural system and the availability of resources. This "positive analysis" seeks to develop a consistent model of the actual development of an economy on a local, regional, sectoral, or global scale. In contrast to this, a "normative analysis" seeks to find an appropriately defined optimal path of economic development which may not be, and more often is not, identical to the actual path.

The Positive Analysis

More formally, let $u_i\{t, h[x(t), e(t)]\}$ denote the utility of an economic agent i (i=1, ...,n) at time t. Utility is determined by the vector of activities $x(t)$ at her disposal and by a vector of resources from nature $e(t)$ which interact to provide a vector of market and non-market commodities and services $h(\cdot)$. The maximization of each agents utility with respect to the activities takes place over each agents time horizon T with an individual rate of time preference r_i:

$$\max_{x_i(t)} \int_t^T u_i\{t, h[x_i(t), e(t)]\}e^{-r_it}.$$

The individual decisions of all agents determine the development of an economy whereby the individual decisions are coordinated by market transactions and government intervention. It is clear that such a path may not be optimal for a number of reasons. At each point in time externalities between agents may not be considered, over time the utility of future generations is not taken into account explicitly. Such interests of future generations may be indirectly represented in the utility function of the agent i through bequest motives. Such an analysis can illuminate the determinants of the emission path of greenhouse gases, it can under the ceteris paribus assumption make conditional predictions about future economic development. These models can also be used to compare the allocation of resources under different institutional regimes – e.g. different climate policy regimes.

The Normative Approach

The normative approach asks a different question, namely: What is an optimal path of economic development for a region, a sector, an economy, or the world economy. Answering this question requires a definition of optimality. Especially for the climate problem as a global phenomenon a global welfare function needs to be determined. This step raises not only modeling problems, it requires ethical and distributional judgments. A global welfare function needs to aggregate welfare of different regions, i.e. it must make comparable the utility of individuals in rich and poor countries. Implicitly, such an approach also assumes that a decrease in utility in one region can be substituted by an increase in another region.[1] The second aggregation needs to make comparable the welfare at different points in time. To do this, one or several discount rates for discounting welfare within and between different generations need to be chosen.

The optimal control problem can then be stated as maximizing the present value of world welfare, i.e.

$$\max_{x(t)} \sum_{i=1}^n \alpha_i u_i(\cdot)e^{-r_i(t)\cdot t}$$

---

[1] A possible alternative would be to assume that the welfare of different agents or regions is not substitutable. But then the optimal path would only consist of strategies which are Pareto-improving. The set of such strategies may well be empty.

where α_i are welfare weights to be chosen and $r_i(t)$ is a possibly time dependent and possibly agent dependent rate of discount of utilities.

In addition to the difficulties involved in determining a world-wide welfare function and in choosing appropriate discount rates[2] the optimal control approach requires a complete linkage to the climate system if it were to derive an optimal long-run development of the economy which takes into consideration the impact of economic decisions on the climate system and the impact of a changing climate on economic activities. Because of the large informational requirements the optimal control approach is used only at a high level of regional and sectoral aggregation and with reduced form linkages to the climate models (see e.g. Nordhaus 1994, Nordhaus and Yang 1996).

Whereas the normative approach asks the question: what would be the best growth path of the world economy given our limited knowledge about the climate economy interaction through emissions and their feedback through the impacts, the positive analysis poses a number of more modest questions such as: what is the likely future impact of climate change on the world economy if one assumes that it continues to develop in its present imperfect development? Or, what would be the economic and climate consequences of specific climate policy options? Such approaches do not require a full linkage of climate models, impact models, and economic growth models and can thus address research questions at a more realistic and more disaggregated level of analysis.

Economic Development and CO_2-Emissions

Today's concentration of greenhouse gases in the atmosphere is the result of economic activity during the 20th century as described by the energy mix in figure 1. It is expected to increase much further in the future as economic activity is expected to expand world wide. For rational policy choices it is necessary to obtain an estimate of the approximate future emissions of greenhouse gases, especially of CO_2, in order to determine how large and how intensive the policy intervention in the energy use needs to be in order to limit emissions to a degree which result in a tolerable extent of climate change for future generations.

Economic development currently takes place at quite different speeds in the world economy and it is expected that this trend of unequal growth will continue for the foreseeable future. This has the practical consequence that today's large emitters of CO_2, namely the industrialized countries, especially the United States, will not be the largest emitters in the future. These positions will be taken by China and possibly India. Consequently, strategies for long-term climate policies need to take into account these changes in the regional composition of economic growth, technological development, and eventually the CO_2-emissions. Another long-run uncertainty concerns the reserves of fossil fuels which in the case of oil by many are expected to start to diminish for the first time in history during the

---

[2] In fact, the choice of a discount rate is the most crucial one if the optimisation takes place over many decades to up to a century since such distant impacts would be discounted to almost zero at even small discount rates.

first decade of this century (Campbell and Laherrère 1998, Kaufmann 1991). Similarly, gas reserves will also become scarce in the next decades, whereas coal will remain in ample supply. The determinants of supply behavior of countries with large reserves of fossil fuels are therefore important.

The last important factor influencing CO_2-emissions is the speed at which technological progress reduces the energy input necessary to produce a given amount of goods and services in the different regions. Currently, the energy efficiency in production differs widely across regions as does the general level of technology. The industrialized countries emit somewhat more than 160 tons of CO_2 per million \$ GDP produced whereas the rest of the world economy needs around 600 tons to produce the same GDP. It is a stylized fact of economic growth that technical progress is faster in low income countries than in economically advanced economies. This is mainly due to technology transfer which is the more pronounced the more an economy is open to trade in goods and services as well as to foreign capital. The critical question is whether this technology transfer is strong enough to mediate the increasing demand for energy from the general catch-up process of the low-income countries.

These three major drivers of economic development, i.e. regional growth rates depending on savings and investment, the supply behavior of the owners of fossil fuel reserves, and the speed of technical progress including technology transfer will determine the future path of the use of fossil fuels and the path of emissions of greenhouse gases of which CO_2 is the dominating one. In each of these three areas one is faced with considerable uncertainties about the future development of key parameters. Regional growth is mainly determined by the amount of saving in the economy but also by many political and institutional factors. Whereas savings have not been fluctuating much in the past, institutional surprises are numerous. By ignoring these political and institutional crises future economic growth can be predicted from past experiences. The uncertainties in the future supply of the fossil fuels coal, gas, and oil as well as their non-fossil substitutes renewables, nuclear energy, etc. is more severe as it depends to an even larger degree on political factors, especially on climate policy. The future energy mix as it has been developed in the different scenarios shown in figure 1 are all plausible paths depending on the level and structure of prices for different energy sources.

It goes without saying that sustainable development does not only mean a stabilization of the climate system but also the achievement of a sustainable level of economic well-being. This requires for the large majority of the world's population an increase in per capita incomes. How much natural and environmental resources are needed to meet this goal depends on the resource efficiency of production which can be achieved in the future. The extent of the conflict between protecting the climate system and raising incomes of the poor is largely determined by the advances in energy efficiency.

Figure 3 summarizes the development of per-capita incomes, of CO_2-emissions, and of the emission intensity (CO_2-emissions per unit of GDP) for the industrialized countries and the Third World including the transition countries up to the year 2030. These results are based on a dynamic, multi-regional, multi-sectoral, computable general equilibrium model (Springer 1999, Klepper and Springer 2000) with 11 world regions and 10 industry sectors. It uses the available

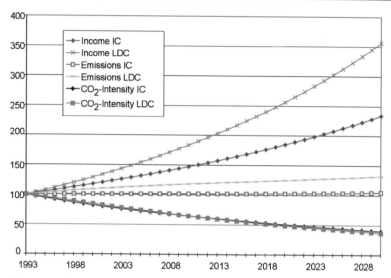

Fig. 3. Development of Per Capita Income, Per Capita CO_2-Emissions, and CO_2-Intensity Industrialized Countries (IC) and Developing Countries (LDC), Index 1993 = 100. [Source: Own calculations (Klepper and Springer 2000).]

empirical information about the drivers of economic development described above. The supply scenario of fossil fuels follows the central case B shown in figure 1 with a roughly constant share of coal and a substitution of crude oil and gas by renewables and nuclear energy.

This scenario shows approximately constant emissions in the industrialized countries at 3 gigatons despite a continuous increase in per-capita incomes. The increase in total emissions will therefore come solely from the developing countries which will grow from roughly 4 gigatons today to about 6.5 gigatons in 2030. Nevertheless, per-capita emissions will only grow from around 700 kgC to 900 kgC mainly because of population growth whereas industrialized country emissions are around 3,700 kgC per capita. These emissions are accompanied by a strong growth in per-capita incomes in the developing countries and somewhat lower growth in the industrialized countries. The CO_2-emission intensity will fall throughout the world. However, today the industrialized countries emit around 160 tons of carbon per million dollars of GDP, whereas the rest of the world needs emissions of 600 tC to produce the same GDP. In this scenario without climate policy intervention CO_2-intensities will fall about 2.1 percent per year in the industrialized countries and about 2.5 percent in the Third World. This exogenous technological progress combined with a restructuring of the energy mix will lower the CO_2-intensities to around 75 tC/mio $ GDP and 230 tC/mio $ GDM. I.e., the intensity of emissions has been cut by more than 50 percent in the industrialized countries and by over 60 percent in the other economies. Nevertheless, economic growth in the Third World is stronger than the improvements in the CO_2-efficiency such that per-capita emissions increase whereas they remain constant in the industrialized countries.

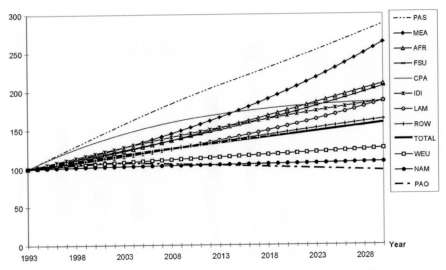

Fig. 4. Regional CO_2-Emissions without Climate Policy, Index 1993 = 100

The expected regional composition of CO_2-emissions is shown in figure 4. The industrialized countries (WEU = Western European Union, NAM = North America, PAO = Japan, Australia, New Zealand) will increase CO_2-emissions by little only. Strong increases are expected in the Asian Tigers (PAS), the former Soviet Union (FSU), the Middle East and North Africa (MEA), China (CPA), and India (IDI). Also increasing are Latin America (LAM), Africa (AFR), and the countries in the rest of the world (ROW).

The increase in the use of fossil fuels is accompanied by an increase in the scarcity of fossil fuels and thus by rising prices which result in higher resource rents for the suppliers of the fossil fuels. Figure 5 shows that in all regions income from supplying fossil fuels will make up a larger share of GDP. Hence, energy will become an even more important factor in the economic system. Whereas the share of income coming from labor will fall that from capital will increase somewhat. This is important to remember since climate policies will reduce the rents from fossil fuels and countries or regions with a high share of resource rents tend to be affected the most by climate policies.

Climate Policies

Despite our limited knowledge about the likely impact of the projected CO_2-emissions on global and regional climate change and despite the long time-lags between emissions and climate change, it seems clear that CO_2-emissions should not increase from today's 6 gigatons to the projected 9–10 gigatons in 2030. Even though the impact of these emissions will only translate into possibly dramatic climate change at the end of the century with unknown economic consequences,

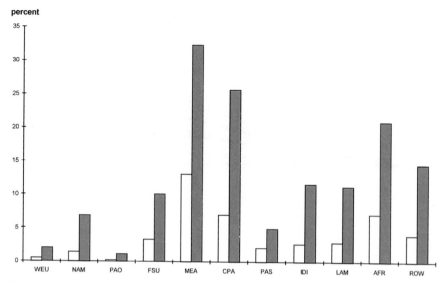

Fig. 5. Share of Rents from Fossil Fuels in Total Income by Region

these impacts will be irreversible within reasonable time horizons. Therefore, a precautionary policy of reducing CO_2-emissions makes good sense. At present, the Kyoto Protocol presents a first step into that direction by setting specific emission limits of greenhouse gases for the Annex I-countries, i.e. all industrialized countries including Russia which are to be met within the period 2008 to 2012. For the other countries no limits are set but offers for reducing emissions through joint implementation and the "Clean Development Mechanism" are made in the Kyoto Protocol.

A comparison of the likely emission path without the limitations given in the Kyoto Protocol and by including its commitments shows that by 2010 CO_2-emissions would have risen to 7 gigatons instead of almost 8 gigatons without climate policy. Without further reduction efforts by all countries the trend to higher emissions will not be reversed. Therefore, the German Advisory Council on Global Change (WBGU) has defined a "tolerable window" of climate change which it has translated in reduction efforts of 3 percent per year for the Annex I-countries of the Kyoto Protocol and a freezing of Non-Annex I-countries emissions at their 2010 levels. These are seen to be necessary to prevent serious negative effects from climate change. This policy scenario will in 2030 reach CO_2-emissions at current levels, however with industrialized countries cutting emissions by 62 percent and the other countries by doubling their emissions.

The Effects of the Kyoto Protocol

The impact of the mild restrictions of the Kyoto Protocol is not only confined to the Annex I-countries. The international division of labor through intensive trade relations has repercussions on the other world regions as well. The restrictions of the Kyoto Protocol improve the competitiveness of Non-Annex I-regions in the production of energy intensive products. As a consequence a relocation of a part of the energy intensive sectors, i.e., energy production as well as the iron and steel industry and the chemicals industry from the Annex I to the Non-Annex I-countries takes place. This so called leakage of emissions, i.e. the increase in emissions due to the relocation of economic activity, rises to about 20 percent. In other words, 20 percent of the Kyoto Commitment will not be effective reduction but will be emitted elsewhere as a reaction of world markets to the new structure of economic activities. The overall adjustment in the structure of the economies to the restrictions of the Kyoto Protocol are summarized in figure 6. It shows that in the Annex I-countries the sectoral changes in production outside the energy sector are very small. An exception is Russia which will be more affected since its economy is heavily oriented towards inefficient and energy intensive activities. The Non-Annex I-regions experience the just mentioned expansion in the energy-intensive sectors; especially China, the Asian NICs and Africa profit. On the other hand, some regions such as the Middle East and the rest of the world loose competitiveness in agriculture and labor-intensive products.

The total effect on welfare for the different regions from fulfilling the reduction commitments of the Kyoto Protocol are shown in figure 7. The loss in welfare to Europe and North America in 2010 when the reduction commitments are assumed to be met will be around 2 percent. Russia will loose around 6 percent; it should be noted, however, that the collapse in the Russian economy in the 1990s is not reflected in the model. In a sense the Russian "hot air", i.e. CO_2-reductions beyond the Kyoto Protocol commitment, is already paid for through the economic decline of the Russian economy instead of a specific climate policy. It is also remarkable that some Non-Annex I-regions gain and some loose through the Kyoto Protocol. Several factors determine winners and losers. First of all, exporters of fossil fuels will be negatively affected by the decline in world demand which is accompanied by falling prices. This case dominates in the Middle East. Conversely, importers of fossil fuels would gain such as India and the Asian NICs. The second positive impact on welfare comes from the gains in competitiveness in the energy intensive sectors especially in China but to a lesser degree in other regions as well. Finally, there is a negative spillover from the industrialized countries: The Kyoto Protocol reduces growth in the Annex I-countries and therefore also has a dampening effect on import demand. Hence, exports of Non-Annex I-countries will suffer from slower growth in the Annex I-countries.

Figure 7 also reveals an interesting insight into the efficiency of climate policies. It is theoretically clear that the reduction commitments of the Kyoto Protocol do not present an efficient climate policy. This would only be reached by a CO_2-tax or tradable emission rights. In order to assess the efficiency loss of the Kyoto Protocol the welfare effects of a CO_2-tax regime have been computed. It was assumed that the same overall CO_2-reduction for Annex I-countries is to be achieved through a CO_2-tax for Annex I-countries only. This economically efficient policy will lower welfare and increase welfare gains vis-à-vis the Kyoto Pro-

Region

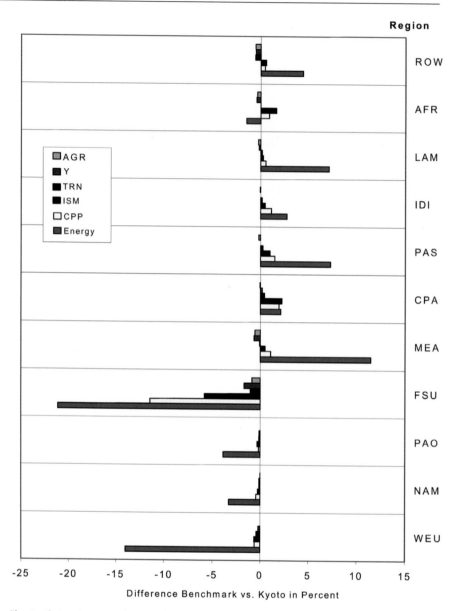

Fig. 6. Change in Sectoral Output in 2010 – Kyoto Protocol
AGR = Agriculture; Y = Labor Intensive Goods; TRN = Transport; ISM = Iron, Steel, Metals;
CPP = Chemicals, Pulp and Paper; Energy = Energy.

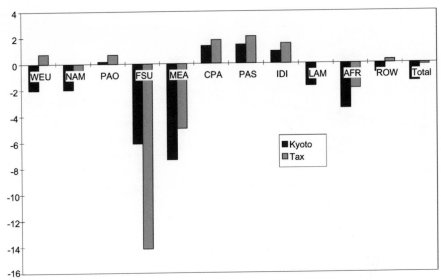

Fig. 7. Welfare Effects of the Kyoto Protocol and a CO_2-Tax for Annex I-Countries (in percent)

tocol with the exception of Russia[3]. The overall world welfare loss under the Kyoto Protocol of almost 1 percent drops to almost zero. There is also some redistribution of gains and losses between Europe (WEU), Japan, Australia, and New Zealand (PAO) on the one hand and North America (NAM) on the other hand. The energy efficiency in the former economies is much higher such that an equal CO_2-tax requires less reduction in energy consumption than in North America.

The Costs of the WBGU's "Tolerable Window"

The Kyoto Protocol will not reduce or even stabilize world wide emissions. Consequently, the comparatively small reduction in emissions has relatively minor welfare losses for Annex I-countries and gains for many Non-Annex I-countries. The more courageous proposal of the WBGU is shown in figure 8. It requires a large reduction of CO_2-emissions in Annex I-countries, 1.5 gigatons instead of almost 6 gigatons in 2030. Non-Annex I-countries are only required to stabilize emissions at the levels reached in 2010. Such large reductions tend to increase the welfare losses.

One can see from figure 9 that already in the period up to 2010 the welfare losses in the industrialized countries increases to around 5 percent compared to less than 2 percent under the Kyoto Protocol. Later the losses increase to around 20 percent relative to the benchmark without CO_2-reductions. The WBGU pro-

---

[3] World-wide efficiency would only be achieved through a world-wide tax. Since this is not accepted as an option, only efficiency within the Annex I-countries is modelled here.

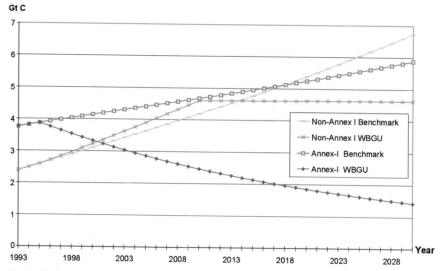

Fig. 8. Carbon Emissions WBGU vs. Benchmark

posal will lead to a strong reduction in demand for fossil fuels such that the major oil exporters – i.e. the Middle East (MEA) – are most affected. Some countries like India (IDI) and the Asian NICs (PAS) still experience welfare gains because the losses due to a reduced energy input are more than compensated by the gains in competitiveness relative to the industrialized countries, i.e. through leakage.

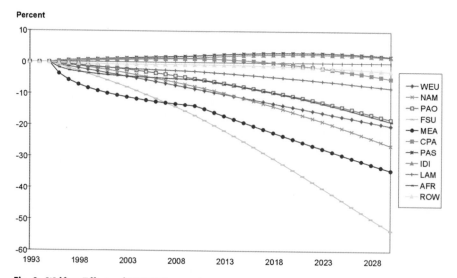

Fig. 9. Welfare Effects of WBGU Proposal

International Coordination of Climate Policy

The simulation exercises of the economic effects of the Kyoto Protocol and of the more ambitions. "Tolerable Window" of the WBGU illustrate several important aspects of the economic effects of CO_2-reduction efforts. The reduction of greenhouse gas emissions in the next decades needs to take into account the likely development of the world economy. Since growth strongly varies between the different economies only a regionally disaggregated view can assess the costs of emissions reductions of particular regions. Furthermore, since the world economy experiences a process of globalization unilateral policies will have multilateral consequences because of the trade and capital market linkages between economies. Therefore, the climate system and its control through climate policy can not be adequately dealt with if it is not linked to a rather detailed analysis of the economic system and its future development. The policy simulations show that the degree of emission reductions by a region is not equivalent to the degree of economic costs incurred by such a policy. The adjustment of the world economy to climate policy initiatives has the effect that the winners and losers are not easily identified. It is a misconception, to expect that economies which reduce CO_2-emissions will lose welfare whereas those without reductions will gain.

Comparing the comparatively soft Kyoto Protocol with the stronger restrictions of the WBGU-proposal shows that the overall cost of climate policy can accumulate to large numbers if the input of fossil fuels is to be cut by large amounts. Such results, however, need to be interpreted with caution. It was argued above that the development of technologies, technology diffusion, as well as the supply behavior of the regions with reserves of oil, gas, and coal are important parameters but difficult to predict. This problem becomes the more severe the stronger the world economy is perturbed by climate policy and the longer the time horizon over which scenarios and policy simulations are run. The WBGU's required CO_2-reductions are of such a large scale that technological changes may be induced which are so far not represented in the simulation model. Especially, non-fossil backstop technologies which are incorporated in many models of economic aspects of climate change as a "free lunch" in the near to distant future are not included in this analysis. In addition, economic models are calibrated to today's economic system with today's set of prices, preferences, and technologies. Large changes in one of these dimensions may turn out to be beyond the usually assumed substitution effects and structural parameters which are commonly assumed to be constant may change instead. Such large scale changes of the economic system still wait for a new type of models.

References

Campbell, C.J., J.H. Laherrère (1998): "The End of Cheap Oil." Scientific American, Vol. 278, No. 3, 60–65.

Kaufmann, R.K., (1991): "Oil Production in the Lower 48 States. Reconciling Curve Fitting and Econometric Models". Resources and Energy, Vol. 13, 111-127.

Klepper, Gernot and Katrin Springer (2000): "Benchmarking the Future. A Dynamic, Multi-Regional, Multi-Sectoral Trade Model for the Analysis of Climate Policies". Kiel Working Paper, No. 976. Institute of World Economics, Kiel.

Nordhaus, William (1994): "Managing the Global Commons: The Economics of Climate Change". MIT Press.

Nordhaus, William and Zili Yang (1996): "A Regional Dynamic General-Equilibrium Model of Alternative Climate Change Strategies". American Economic Review, Vol. 86, No. 4, 741-765.

Springer, Katrin (1999): "Climate Policy and the Steady-State Assumption in a Multi-Regional Framework". Kiel Working Paper 952. Institute of World Economics, Kiel.

CHAPTER 13

From Nature-Dominated to Human-Dominated Environmental Changes

13

Bruno Messerli* · Martin Grosjean · Thomas Hofer · Lautaro Núñez · Christian Pfister

Extended Summary[1]

At this critical moment of Earth's history, as we move from a century with rapidly growing human impacts on all the different ecosystems of our planet, to a century with a probable further acceleration in the pace of environmental change, resource use, and vulnerability for societies and economies, we have to rethink the changing relationship between nature and human beings from the past, through the present, towards a future full of uncertainty. It becomes more and more evident that major natural processes from the local to the global level are influenced by human activities, creating a much higher degree of complexity through the interaction of processes which are within the domain of both the natural and social sciences. This implies a need to bridge the gulf between the two cultures of science in order to advance our understanding of contemporary driving forces and their rapidly growing impact on earth's ecosystems. "The biggest changes happened in our century, more precisely in the last 50 years, with a rate unknown before in Earth's history" (Pfister 1995a). In light of continued population growth, economic development, urbanisation, industrialisation and resource use, it is clear that human impacts on ecosystems world-wide will continue to increase in the next century.

In three sections, representing three case studies in very different time periods (fully documented in Messerli et al. 2000), we try to identify the changing priorities in the nature-human interactions with the following hypothesis: that for any human society there is a historical 'trajectory of vulnerability', beginning with highly vulnerable societies of hunters and gatherers, passing through periods with less vulnerable, well-buffered and highly productive agrarian-urban societies to a world with regions of extreme overpopulation and overuse of life-support systems, so that vulnerability to environmental changes and extreme events is again increasing?

---

* e-mail: messerli@giub.uni-be.ch
[1] The extended summary is a short version of three case studies, which are fully published and documented in: Messerli B, Grosjean M, Hofer T, Nunez L, Pfister Chr (2000): From nature-dominated to human-dominated environmental changes. In: Past global changes and their significance for the future. Guest editors KD Alverson, F Oldfield and RS Bradley. PAGES-Programme. Quaternary Science Reviews, Pergamon, Vol 19 No. 1-5, 459–479.

This question can not be answered through a global approach, because the state of our knowledge is quite insufficient and every region has its own unique natural and cultural history. This richness of highly diversified natural-human interactions constitutes an urgent challenge for interdisciplinary science, because we need a much more careful analysis of the past to understand the present and to develop scenarios for a future in which political-economic factors and cultural perceptions of the environment will be very important driving forces.

Three case studies

– *The Atacama desert – Early-mid Holocene: Nature dominated environment – human adaptation, mitigation, and migration.*
 In the central Andes the Holocene climate changed from more humid conditions (10,800 – 8,000 BP) during the first prehistoric hunting communities in this area to extreme arid conditions (8,000 – 3,600 BP). Water and vegetation became so scarce that these hunting and gathering societies were forced to migrate. Until now, the period between 8,000 BP and approximately 4,000 BP was known in the archaeological literature about the Atacama desert as "silencio arqueológico". Only recently was the evidence for a human survival close to a few perennial water bodies found in the gorge of Puripica, close to San Pedro de Atacama (see fig. 1 and 2). The analysis of the excavated horizons and materials indicated that the process of domesticating camelids began around 5,000 BP and irrigation at about 3,100 BP (see fig. 2). Harsh environmental conditions and water stress in a region with marginal natural resources have resulted in relatively widespread depopulation of an area, and/or triggered adaptive processes (including technological innovations). It is not astonishing that re-occupation between 4,000 and 3,600 BP, coincided with rising lake levels and a marked shift towards more favourable climatic conditions (Grosjean et al. 1997). This case study shows a fascinating process of migration, adaptation and mitigation to reduce the vulnerability to climate change and limited resource availability. The nature-dominated environmental change is evident. However, the higher the degree of complexity in a given society, the harder it is to establish the extent to which certain innovations in technology and practices in resource use were a response to the impact of exclusively natural processes, or reflected a stimulus arising from within the human culture prevailing at the time.
– *Western Europe – historical period: An agrarian society in transition from an "enduring" to an innovative human response*
 Detailed documentary evidence from Western Europe may be used to reconstruct quite precisely the impacts of climatic variations on agrarian societies. The period considered spans a major transition from an apparently passive response to the vagaries of the environment during the 16th century to an active and innovative attitude from the onset of the agrarian revolution in the late 18th century through to the present day. The associated changes in technology and in agricultural practices helped to create a society better able to survive the impact of climatic extremes.

Fig. 1. Rio Puripica (San Pedro de Atacama, Chile).

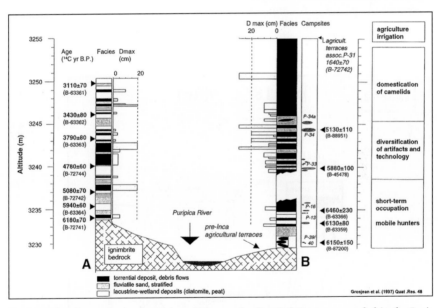

Fig. 2. Mid Holocene environmental conditions and cultural changes as recorded in the Puripica site. Profile A contains the record of individual debris flows (Dmax is the maximum diameter of transported particles). Wetland deposits provide evidence of relatively favourable environmental conditions in the local ecological refuge between 6,200 and 3,100 ^{14}C yr BP, whereas extremely arid and hostile conditions prevailed in the surrounding areas. Profile B shows the archaeological sites embedded in the debris flow deposits. These sites provide information at a very high resolution about a stepwise adaptive process from hunting and gathering societies to complex societies with agriculture and irrigation. The pre-Inca terraces near the modern river bed suggest that the entire sediment sequence was eroded within about 2,000 years.

With the advent of the agrarian revolution from the late eighteenth century, the carrying capacity of the land achieved a quantum leap forward (see fig. 3). Through innovations in agriculture, such as the introduction of leguminous fodder plants (e.g., clover), as well as the building of underground reservoirs to recycle liquid manure and the keeping of cattle in stables throughout the year, new sources of nitrogen were tapped. This allowed a considerably greater extension of the area on which potatoes were grown (Pfister 1990). It must be emphasised that this collection of interrelated innovations was fundamental in promoting the take-off of agriculture which resulted in being able to feed the population even better, despite its rapid growth. Fluctuations in the carrying capacity of the agricultural system no longer triggered immediate demographic consequences. In this sense it was a decisive element in the final success of the industrial revolution in England and in central Europe (Buchheim 1994).

From the mid-twentieth century, the application of (chemical) fertilisers, the use of pesticides and the introduction of high-yield varieties of crops made possible a pronounced increase in yields per acre. But agriculture, combustion of fossil fuels, and other human activities have altered the present day global cycle of nitrogen substantially over large regions of the Earth. What was a success in the past, has become a threat today (Vitousek et al. 1997a). The crea-

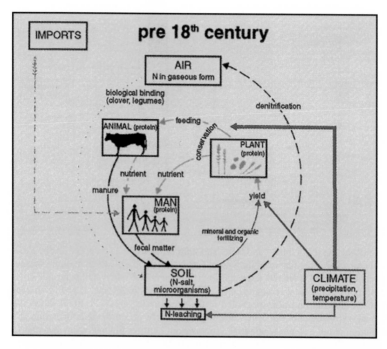

Fig. 3. The evolution of the Nitrogen-Cycle in Central Europe. [Source: Pfister (1984, 1990).]

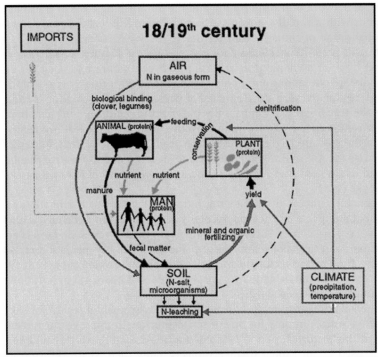

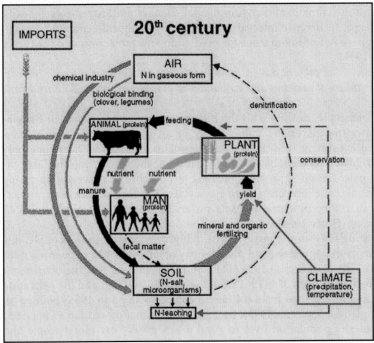

Fig. 3. *Continued*

tion of a global network of transportation and the diminishing costs of transportation together with a rapidly growing urbanisation and industrialisation initiated the globalisation of climatic risk over the last 50 years.

The historical period considered above shows a clear picture, distinctive in space and time, of a changing relationship between nature and societies or between climatic variations and human reactions. Even if it is a very complicated and regionally very differentiated process, it is a fascinating period marking the transition from apparently passive and 'enduring' behaviour in the earlier centuries to an active and innovative attitude in the last two centuries. With the so-called agrarian revolution from the 18th century onwards, stimulated by the Age of Enlightenment, came an active and innovative change of the land use system itself, which created a more efficient buffer against climatic shocks (see fig. 3).

This fascinating process of a changing human relationship to climatic and environmental impacts has become recognisable only since the historical and social sciences have come to accept this important field of research. Moreover, the information in terms of years, seasons, and even daily events is so precise that a correlation between climatic impacts and human reactions, that is to say between semi-quantitative and cross-checked data and historical descriptions and documents, can be established.

A fundamental goal remains the extension of the instrumental record for several hundred years, not only for one or two centuries. To improve our understanding of this historical period is to open a treasure store which could and should be used by all Global Change projects. Only then, shall we begin to understand the climatic and environmental changes and extreme events with their different recurrence intervals in their human-cultural-economic context. This is of paramount importance for prediction and projection into the future (Bradley 1999).

– *Bangladesh – The present day: A human dominated environment with increasing vulnerability of societies and economies to extreme events and natural variability.*

The third example, dealing with the history and impact of floods in Bangladesh, shows the increasing vulnerability of an over-exploited and human-dominated ecosystem (see fig. 4). Measurements exist for a short time only (decades), historical data allow a prolongation of the record into the last century, and paleo-research provides the long term record of processes operating over millennia. The long term paleo-perspective is essential for a better understanding of future potential impacts on an increasingly human-dominated environment (Hofer 1998).

Concentrating only on the present time, Bangladesh is a unique example, due to the density of population and the intensity of land use. The population doubled from 1961 (55 Mio.) to 1991 (111 Mio.). This implies that the density per km^2 has increased from 374 (1961) to more than 800 (1994), and the per capita land holding diminished from 0.40 ha in 1950 to 0.16 ha in 1985 (Elahi et al. 1992, Bangladesh Statistics 1993). Bearing in mind that more than 83 % (1993) of Bangladesh's population lives in rural areas, every year more people and more small holders are affected adversely by floods.

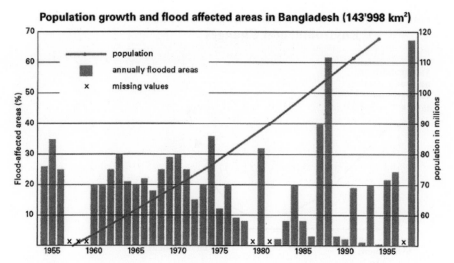

Fig. 4. Population growth (Bangladesh Statistics, 1993) and flood affected areas Source: BWDB (1991), FPCO (1995).

The example of Bangladesh may show how human occupation and domination of a certain type of environment has gone beyond the limits of long term viability. The threshold is dictated by nature, but the interaction between nature and human activities is a highly dynamic one. The increasing human population was forced to change land cover and intensify land use in an endangered area in such a way that the risk for loss of life and property has increased dramatically. More precisely: the use of former swamps and lakes and the interruption of former feeder channels by constructions, roads, embankments, etc. has led to a loss of storage capacity for flood water for the Brahmaputra from Assam to Bangladesh. This means that the rapidly growing population, through its own activities, has increased its own vulnerability, against any advice based on past experience, but in response to the immediate need to survive without an opportunity to migrate from the country. Without discussing the causes of the floods and the proposed technical solutions (Hofer and Messerli 1997, Hofer 1998), we may state: Bangladesh is an alarming example of a rapid transition from a nature-dominated to a human-dominated environment, and exactly for this reason it becomes again highly vulnerable to nature-dominated processes (see fig. 5).

Similar examples from recent years could be drawn from developing and from industrialised countries, e.g. Mississippi, Yangtse, Rhine, etc. Flood plains are especially suitable for intense land use, urbanisation, industrialisation, traffic systems, etc. It is precisely in these regions that we see the most dramatic combination of a human-dominated environment and a high vulnerability to extreme climatic-hydrological events.

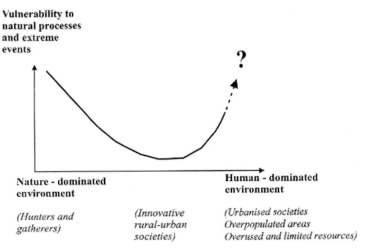

From Nature - Dominated to Human - Dominated Environmental Changes

Fig. 5. Vulnerability of human societies to natural processes and extreme events from the past to the future. Are we moving towards a higher vulnerability in certain areas of the "developed" and the developing world? Should the so-called critical regions not have a higher priority in our research agenda?

Conclusions

Even if we could place all three case studies on a time-axis, it would be completely wrong to generalise to the point of looking for comparable processes globally over the same time intervals:

- Let us take the first case study in the time period between 5,000–4,000 BP as an example: in the Atacama region began a society of hunters with the domestication of camelids. In the same period we had the so-called "golden age" in Egypt with the construction of the pyramids, the development of a productive irrigation based agriculture and a high level in arts and crafts.
- In the second case study, during the historical period, the development towards a modern agrarian society, as described here, is typical for Europe with its particular environmental conditions and its cultural history. Even if we can generalise some elements of this process, due to the worldwide impact of the European innovations, we should bear in mind that this process must have been and will be very different in other parts of the world.
- Bangladesh, as the third case study, is certainly an extreme example with endangered life-support systems, high risks and increasing vulnerability. Even if we can generalise only some few elements of interaction between natural processes and human systems, it may give us at least an idea what could happen in the future in other regions of the world.

These observations may serve to indicate the spatial and cultural diversity of the processes with which we are concerned. They also highlight the uncertainties in our present state of knowledge, not only on the natural science side, but also on the social-cultural science side as we seek for a better understanding of the processes involved in any transitions from nature-dominated to human-dominated environmental changes in all their spatial and temporal complexity.

At the end of this century, we are beginning to understand for the first time the degree of human domination of the Earth's ecosystems (Vitousek et al. 1997b) and we begin to realise the value of these ecosystems (Costanza et al. 1997), with all their consequences for a new social contract for science (Lubchenco 1998). We begin to understand the dominant role of human activity in our environment and its consequences for natural systems and natural resources. This conclusion is not limited to developing countries. Higher densities of human population in a growing number of megacities and in endangered rural regions, but also in areas with high pollution, limited resources and endangered life support systems will lead to higher vulnerability and higher risks, to health problems, to loss of life, property and expensive infrastructure, even if modern technology has helped to create a certain buffering capacity against all types of hazards (see fig. 5). This inevitable development towards an even stronger human domination of Earth's ecosystems will need also an ever better knowledge of past events ranging from longer term climate changes to abrupt shifts in régime, to all types of extreme events, so that planning can be undertaken in the knowledge of the full range of relevant past variability.

The more human actions modify, dominate or even replace natural ecosystem, the more they can increase the vulnerability to climate changes or extreme events (see fig. 5). The integration of all these interacting human and natural forcings becomes very complex and perhaps we have to ask ourselves if the traditional ways of organising research are still functional, especially for addressing the most pressing problems of the next century. In the context of the research fields relevant to the study of global change, we should keep in mind four points:

- The human dimension: The growing impact on Earth's ecosystems is very difficult to analyse in all its quantitative and qualitative components, but it is most important to realise that different demographic, cultural, economic, etc. forces must have very different effects; they can be neither ignored, nor over-generalised.
- The regional dimension: We need, besides the global approach, also a strongly regional approach: for example, the Holocene arid and humid periods in the Atacama and in the Sahara did not coincide. The historical period in Europe and its environmental impact were very different from the course of events in other continents. The hydrological processes in Bangladesh are very different from those in arid and semi-arid regions. Without a better understanding of regional, sometimes even local processes, we shall never make the required improvements in global change research, especially those that are essential for improving our capacity for projections into the future (see also Dirzo and Fellous 1998).
- The temporal dimension: without an extension of the all too brief timeframe of instrumental measurements, we shall base our research on insufficient or

misleading information. The last 100–200 years will never reveal the full range of extreme climatic intervals or events which we need to document and understand for any realistic projection into the future. We have to integrate historical data and the results of the different paleosciences in our time series: "The present without a past has no future" (Geoparks 1999).

- The natural dimension: understanding natural processes and their variability from the regional to the global level and on different time scales is fundamental for our future. Climatic and environmental catastrophes are only catastrophes because human beings and activities are involved. For nature alone, climate changes, floods, earthquakes, volcanic eruptions, etc. are a self-evident part of its dynamic processes. Ignoring this natural system leads to increased vulnerability and more frequent catastrophes. Economic growth has natural limits. Until now, increasing human activities have altered nature. In the new century nature could alter humanity. Maybe it could become true what Nobel Price Winner Jody Williams in 1999 said: "Isn't it rather strange that the biggest risk facing mankind is mankind itself?".

References

Bangladesh Statistics (1993) Statistical Pocketbook of Bangladesh. Bureau of Statistics, Dhaka

Bradley RS (1999) Paleoclimatology. Reconstructing Climate of the Quaternary. Internat Geophysics Series 64: 613 pp

Buchheim C (1994) Industrielle Revolutionen. Langfristige Wirtschaftsentwicklung in Grossbritannien, Europa und in Übersee. DTV, München

BWDB (1991) Bangladesh Water Development Board. Dhaka (raw data)

Costanza R, Arge R, Gruot de R, Farber S, Grasso M, Hannon B, Limburg K, Naeem S, O'Neill RV, Paruelo J, Raskin RG, Sutton P, Belt van den M (1997) The value of the world's ecosystems services and natural capital. Nature 387: 253–261

Dirzo R, Fellous JL (1998) Strengthening the regional emphasis of IGBP. Global Change Newsletter 36: 5–6

Elahi M, Sharif AHM, Kalam AKMA (eds) (1992) Bangladesh: Geography, environment and development. National Geographical Association. Dhaka, 171 pp

FPCO (1995) Bangladesh Flood Action Plan. Progress report. Ministry of Water Resources. Dhaka, 46 pp

Geoparks (1999) UNESCO network of Geoparks (Quotation of F.Broudel, French historian, 1902–1985). UNESCO. Division of Earth Sciences, Paris, 6 pp

Grosjean M, Núñez LA, Cartajena I, Messerli B (1997) Mid-Holocene Climate and Culture Change in the Atacama Desert, Northern Chile. Quaternary Research 48: 239–246

Hofer T (1998) Floods in Bangladesh. A Highland-Lowland-Interaction? Univ. of Bern, Geographica Bernensia G48, 171 pp

Hofer T, Messerli B (1997) Floods in Bangladesh. A Synthesis Paper. Univ. of Bern, 32 pp

Lubchenco J (1998) Entering the Century of the Environment: A new Social Contract for Science. Science 279: 491–497

Messerli B, Grosjean M, Hofer T, Nuñez L, Pfister Chr (2000) From nature-dominated to human-dominated environmental changes. In: Past global changes and their significance for the future. Guest editors Alverson KD, Oldfield F, Bradley RS. PAGES-Programme. Quaternary Science Reviews, Pergamon 19 (1–5): 459–479

Pfister C (1984) Klimageschichte der Schweiz 1525–1860. Das Klima der Schweiz und seine Bedeutung in der Geschichte von Bevölkerung und Landwirtschaft. Paul Haupt, Bern

Pfister C (1990) The Early Loss of Ecological Stability in an Agrarian Region: Nitrogen, Energy and the Take-off of 'Solar' Agriculture in the Canton of Bern. In: Brimblecombe P, Pfister C (eds) The Silent Countdown. Essays in European Environmental History 36–55. Springer, Berlin

Pfister C (ed) (1995a) Das 1950er Syndrom. Der Weg in die Konsumgesellschaft. University of Bern, Haupt, 428 pp

Pfister C (1995b) The Population of Late Medieval and Early Modern Germany. In: Scribner B (ed) Germany: A New Social and Economic History, 1450–1630, 1, 33–62. Arnold, London

Vitousek PM, Aber JD, Howwarth RW, Likens GE, Matson PA, Schindler DW, Schlesinger WH, Tilman DG (1997a) Human Alteration of the Global Nitrogen Cycle: Sources and Consequences. Ecological Applications, Ecological Soc. of America 7 (3): 737–750

Vitousek PM, Mooney HA, Lubchenco J, Melillo JM (1997b) Human Domination of Earth's Ecosystems. Science 277: 494

Williams J (1999) On the threshold of the 21st century. Win Conference 1999, Interlaken. "Winterthur", Swiss Insurance Company

Appendix:
Working Group Reports

CHAPTER 14

Water Quality and Health Risks

14

THOMAS KISTEMANN* · MARTIN EXNER*

The decrease of death as a result of infectious diseases in late 19th century Europe was strongly associated with the sanitation of the cities, namely the implementation of sewage systems and water treatment by sand filtration and disinfection. The availability of water is a prerequisite for improved health and sustainable development. It has social, economic and environmental values. Today, there is no doubt that the quality of water for human consumption is of utmost importance for Public Health.

The World Health Organizations regional office for Europe assigned this in 1999 by passing a Protocol on Water and Health (UN/ECE and WHO/EURO 1999). The main targets are:

- Access to drinking water for everyone
- Provision of sanitation for everyone.

This contrasts with the fact that world-wide one billion of people are without access to clean water, and three billions have no access to sanitation infrastructure.

In taking measures to implement the WHO Protocol, the following principles and approaches shall be considered:

- the precautionary principle
- the preventive action principle
- the polluter-pays principle
- management of water resources with regard to the needs of future generations
- management of water resources at the lowest appropriate administrative level
- management of water so as to realise the most acceptable and sustainable combination of its social, economic and environmental values
- management of water resources in an integrated manner on the basis of catchment areas, with the aims of linking social and economic development to the protection of natural ecosystems
- promotion of efficient use of water through economic instruments and awareness-building
- equitable access to water for all members of populations
- special consideration of the protection of people who are particularly vulnerable to water-related disease

* e-mail: boxman@ukb.uni-bonn.de

- installation of comprehensive local surveillance and early-warning systems to identify outbreaks or incidents of water-related disease
- contribution of persons and institutions to the protection of the water environment and the conservation of water resources
- due account to local problems, needs and knowledge
- public access to information concerning water and health
- public awareness for the relationship between water management and public health.

The Break-Out Group "Water Quality and Health Risks" took the WHO Protocol as a framework and discussed the microbial and chemical burden of water for human consumption, the use of GIS-applications in the field of water and health and the economic perspective of water related infectious diseases. Starting from the European view, world-wide consequences and perspectives were discussed. It turned out clearly that there is an urgent need of a holistic approach in water hygiene, comprising methods of prevention (quality of structure and process, quality assurance) and methods of control (surveillance, incident and outbreak management) (see table 1).

Table 1. The basic concept of water hygiene

Water hygiene				
Prevention		Health protection	Control	
Quality of structure and process	Quality assurance	availability of water sufficient in quantity and quality, with special regard to waterborne pathogens: for the public including ill and immuno-suppressed people for food preparation	Surveillance	Incident and Outbreak Management
conservation of water resources catchment areas quality of raw water water treatment disinfection distribution system plumbing system water outlet	risk analysis investigation of water quality indicators of water quality standard methods responsibility education training research standards legislation		collection and analysis of data early-warning systems management of data responsibility education training legislation information of the public	contingence plans for response systems capacity to respond to and to investigate the causes documentation and information system lessons for the future (to prevent incidents and outbreaks)

Water related health risks from microbial burden

Today, microbial pathogens are regarded to be the most important water related health risk factors. World-wide, waterborne infections contribute essentially to diarrhoea diseases, which claim about two million lives per year among children younger than five years. Emerging pathogens (cryptosporidium, E-coli O157:H7, Helicobacter pylori, Legionella) present an additional huge health and economic burden. Outbreaks of waterborne diseases (Cholera in Latin America, 1991: Exner and Böllert 1991; Cryptosporidiosis in the U.S.A., 1993: MacKenzie et al. 1994) have had dramatic health consequences.

There are still many problems to solve in the developed countries, but for most developing countries waterborne diseases are a dramatic impact. In India[1], for example, more than 200 million people do not have access to safe drinking water. Nearly 143,000 villages still have acute water problems. Increasing demands due to population and industrial growth and agricultural development pose new challenges. 75 percent of the waste water is from the domestic sector, but sewage treatment facilities are inadequate in most cities and almost absent in rural India: Only 25 percent of class I cities (i.e. population > 100,000), and less than ten percent of the smaller cities have wastewater collection systems. About 580 million people do not have access to toilets of any type. Fast and unbalanced urbanisation, fed both by massive migration and population explosion, deforestation, improper solid and liquid waste disposal and lack of safe drinking water are the factors which facilitate to create an environment conductive to water borne infections.

As a consequence, 1.5 million children younger than five years die each year due to water borne diseases in India. The high incidence of mortality and morbidity rates among infants and children is attributed largely due to unsafe water supply, poor hygienic practices and insanitary environment. Improvements of sanitation facilities and water supply would reduce death due to diarrhoea by 65 percent and overall child mortality by 55 percent.

Water related health risks from chemical burden[2]

Harmful chemicals and substances in drinking water may originate from anthropogenic emissions (pesticides, nitrate, petrol products, heavy metals), biological processes (toxins from cyanobacteria), geogenic minerals (arsenic, fluoride, for instance), or combined processes (mobilization of heavy metals into groundwater following anaerobic bacterial denitrification of nitrate and oxidation of sulphides).

---

[1] The following sequence is based on commentaries by Prof. Dr. Abha Lakshmi Singh (Aligarh Muslim University, Aligarh/India) and Prof. Dr. S. Shanmuganandan (Madurai Kamaraj University, Palkalaingar/India) within the breakout group.

[2] This chapter is based on the presentation of Prof. Dr. Hermann H. Dieter (Federal Environment Agency, Berlin/Germany) within the breakout group.

Methemoglobinemia following intake of *nitrate* with drinking water is the most serious acute health problem in many states of Eastern Central Europe and in developing countries. High natural *arsenic* concentrations with concentrations of 100 µg/l and much more are a problem in Hungary and Romania, and since about ten years also in West Bengal (Saha et al. 1999). *Fluoride* poisoning by technical failure of a fluoride dosing system arrived in Hooper Bay, Alaska (Gessner et al. 1994). Geogenic fluoride in drinking water poisoned at least 58 children in the City of San Luis Potosi in Mexico (Grimaldo et al. 1995). Concern for Europe's Tomorrow (WHO 1995) quotes a ministerial report of the Czech Republic listing exposure of 553,000 people from 123 localities to *heavy metals*, 10,000 people from ten localities to cyanides, 230,000 people from 57 localities to *phenolics*, one million people from 169 localities to *oil substances*, and 300,000 people from 100 localities to enhanced *radioactivity*.

A further health issue is cancer risk from by-products of chlorination. *Chloroform* and *bromoform* are carcinogenic to animals at high doses, other *trihalomethanes* are suspected to be mutagenic. In several cities of Ukraine, concentrations of several 100 µg/l were detected in chlorine-treated drinking water, which was abstracted from surface water rich in organic precursors (Hoffmann 1995).

Cyanobacteria (blue-green algae) produce a variety of toxins linked to serious health effects (hepatotoxicity, neurotoxicity, cancer). Moreover, they impair drinking water disinfection and are a substrate for bacterial growth in drinking water mains.

Pollution of drinking water by waste water is the main reason for health hazards world-wide. In China, about 85 percent of all cases of illness associated with chemical contamination of water supplies are linked to polluting by sewage. Enhanced rates of cancer mortality have been tied to the lack of adequate sewage treatment. For instance, acute water shortages cause industrial wastewater to be used for agricultural irrigation without treatment (Wu et al. 1999). Therefore, improvements of sewerage and wastewater facilities must always be part of the efforts to improve drinking water supply.

Source protection instead of over-regulating the quality of the end product appears to be the most effective approach to get safe drinking water. Where lack of water due to pollution of sources and their restoration is feasible, re-infiltration of pre-purified waste water into the original resource should be the first option. Bank filtration of river water has principally proven to be a very suitable method to improve water quality. If this is not possible for geological reasons, artificial recharging of groundwater with pre-purified waste water may be carried out nearby in more suitable areas. If groundwater is not available, surface water has to be used directly, but treated adequately by sand filtration, oxidation, flocculation, filtration on activated carbon and disinfection.

Drinking water should be available constantly, not interrupted by urgencies in demand, pollution or breakdown. If the proper supply is not sufficient, people will seek for alternatives which tend to be of poorer quality.

An introduction into the derivation of health-based guide values for contaminants reaching drinking water despite effective emission control measures is given by WHO (1993). The next step should be to apply some principles of environmental hygiene to get and maintain drinking water sufficient and suitable for human consumption (Dieter 1998):

- to avoid useless loads (e.g. pesticides, nitrate), to minimise the functional ones and their by products (e.g. trihalomethanes; metals from corrosion of drinking water pipes), and to prevent those which directly entrain health risks to prevent pollution at its source: this is at a long range more economic and safer than elimination of pollutants from the water;
- to consider adequately treated and filtered waste water as an important drinking water resource.
- to lead back used and pre-purified ground water into the soil close to where it has been abstracted rather than discharging and diluting it into rivers and oceans;
- to make drinking water treatment as simple as possible but as sophisticated as necessary. The purer the raw water, the fewer treatment steps are required and the lower is the risk of technical failure.

Managing water related health risks: the GIS approach

Some principles of the WHO's protocol deal with problems around peace and spatial information:
- What is happening in the catchment area?
- Where have cases of waterborne disease been detected?
- Is there a spatial pattern which might be interpreted as an outbreak?
- Does any spatial relation exist between water management and public health?
- How can information concerning water and health be presented to the public?

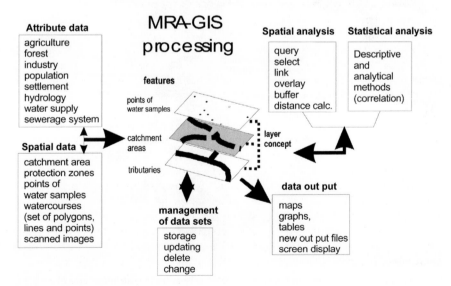

Fig. 1. MRA-GIS-concept

Peace is central to all these questions. Huge quantities of information are available today, far more than ever before. Data abounds from thematic mapping of catchment areas, plans of water-pipes, but also from billing records and customers complaints, from cases of notifiable infectious diseases and health insurance statistics.

To manage spatial data, Geographical Information Systems (GIS) have been developed during the last decades as a new database concept. Recent advances in technology and increased awareness have created new opportunities for public health administrators to enhance their planning, analysis and monitoring capabilities.

A GIS is an organised collection of computer hardware, software, geographical data and personnel designed to efficiently capture, store, update, manipulate, analyse and display all forms of geographically referenced information (WHO

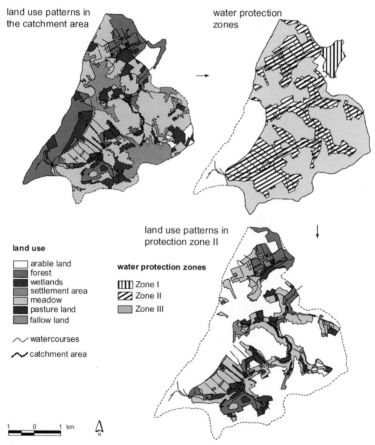

Fig. 2. Overlaying the land use and the protection zone layer for the catchment area of a surface water reservoir (Kall/Germany)

1999). It is an information system with a geographical variable which enables users to easily process, visualise and analyse their data or information spatially. Each piece of information is related in the system through specific geographical co-ordinates to a geographical context. GIS can be applied to support microbial risk assessment (MRA) of water supply systems, for instance (see fig. 1).

As with all databases, GIS is only as powerful as the raw data allows. It is therefore essential that the data are of the highest quality. Although some datasets can be purchased, data requirements are often specific to the tasks. This necessitates the generation of customised datasets.

Current applications of GIS in the field of water and health comprise management of catchment areas and administration of distribution systems as well as support of epidemiological purposes and risk communication (North West Water (NWW) 1999).

In context of monitoring water samples for microbial contamination a detailed *ecological characterization of catchment areas* of several drinking water reservoirs was carried out (Kistemann et al. 1998). The objective was to reveal the origin of microbial loads present in the watercourses. Information about agricultural land use, land cover, population density, settlement areas, sewage systems, etc. were collected, stored and analysed in a GIS. Different GIS-tools helped to deal with

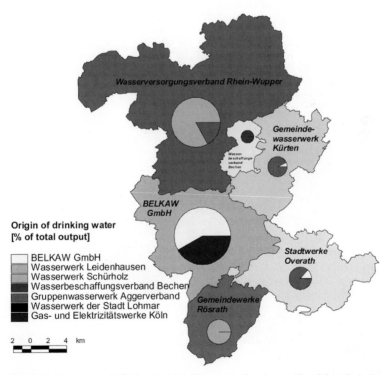

Fig. 3. Production and purchase of drinking water for the supply of the Rhein-Berg district in 1996 (Germany)

several questions. Overlaying the land use and the protection zone layer allowed to calculate the amount of different land use in the protection zones II and III, for instance (see fig. 2). The change of land use proportions from zone to zone differed substantially for different drinking water reservoirs.

For epidemiological purposes, it is very important to know the *patterns of water supply*. In a study area (Rhein-Berg) of 440 km² and 270,000 inhabitants near Cologne, a wide range of data about water supply has been collected concerning origin and quality of the raw water, water processing facilities, interchange of drinking water between water works, quality of the finished water, geometry and material of the grid of water pipes, location of private wells, etc. The aim was to build up a comprehensive drinking water information system, being a fast source for information during incidents of possibly water related disease (see fig. 3).

One major area in which GIS and health research have come together is via the study of environmental and geographical epidemiology. Links between disease

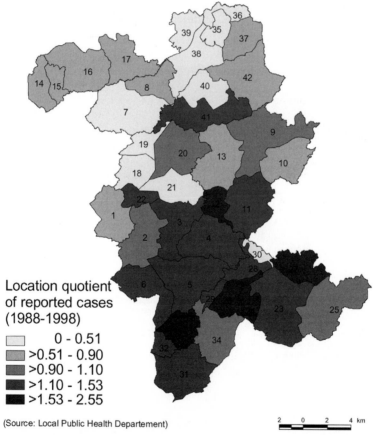

Location quotient of reported cases (1988-1998)

☐ 0 - 0.51
☐ >0.51 - 0.90
☐ >0.90 - 1.10
☐ >1.10 - 1.53
☐ >1.53 - 2.55

(Source: Local Public Health Departement)

Fig. 4. Location quotient of enteritis infectiosa-cases in the Rhein-Berg district (Germany)

and environment, disease clustering, cluster identification, associations with potential point and line sources of pollution, and of space-time disease incidence, are investigated (Dunn 1992, Clarke et al. 1996). GIS is used in the surveillance and monitoring of waterborne diseases, and especially to support outbreak investigations (NWW 1999).

For the Rhein-Berg study area address-based epidemiological data of gastrointestinal infections were available from the local public health department for the period 1988–1998. GIS provided a linkage between address-coordinates and cases. It is possible to display

- point patterns of cases for different periods to show trends
- case specific queries about the distribution of diarrhoea diseases in the study area concerning date of infection, pathogen, suspicious source of infection etc.

Aggregating the data on the smallest administrative units results in choropleth maps of incidence, which can be tested for heterogeneity by spatial autocorrelation tests. Location quotients, which are the quotient of observed and expected value for every unit, make interpretation much more straightforward (see fig. 4). The next step is to perform an ecological study and test for associations between the patterns of incidence rates and of interesting environmental factors.

Economic Assessment: A European Perspective[3]

Many economic consequences follow from the quality of water and sanitation infrastructure a country is able to provide for its citizens. Improvements in the level of service will lead to improvements in health but also involve a cost as the funds devoted to water and sanitation improvements are not available for other purposes. Ignorance of the ways in which inadequate water supply and sanitation systems lead to health and other costs, increases the risk of inadequate investment.

Across the wider European region there is a stark contrast between the quality of water and sanitation facilities, and one consequence of programmes to control and prevent water related disease across the region would be a narrowing of the gap between West and East. The role of the economist is to enumerate as far as possible the costs and the benefits of programmes, which secure environmental improvements.

An economic assessment of water related health impacts must examine and build four relationships between water, health and economics:

- Expenditure on water supply/sanitation and contaminant exposure
- Contaminant exposure and disease incidence
- Disease incidence and costs of illness
- Costs of illness and expenditure on water supply/sanitation.

---

[3] This chapter is based on the presentation of Kevin Andrews MA MSc (Water Research Centre, UK) within the breakout group, which was a summary of a study undertaken by him in support of the joint World Health Organization/European Environment Agency Monograph on Water and Health in Europe (WHO/EURO,EEA 1999).

Within the context of an economic assessment of programmes to control and prevent water related diseases the forth relationship is of primary interest. But the shape of this relationship depends fundamentally on the nature of the other relationships within the system.

Even for the European region, the available information on the incidence of water related diseases is extremely limited. Data had to be interpolated, starting from data of recorded cases of various water related diseases (cholera, typhoid, amoebic dysentery, hepatitis A, giardiasis, bacillary dysentery, general gastroenteritis) which were collected recently through a questionnaire sent to countries as part of the WHO/EEA Draft Monograph on Water Resources and Health (WHO/EEA 1999). Furthermore, the amount of cases which are water related and are avoidable through investments in water supply and sanitation had to be estimated. Between 68 and 77 % of the reported cases were assumed to be water related and avoidable. Under-reporting of diseases was accounted for by extrapolating from the reported cases to the total number of cases. Evidence from the west suggested that the level of underreporting for non-severe diseases (giardiasis, bacillary dysentery, general gastroenteritis) is in the region of 30 to 1. Underreporting was not assumed for the more severe diseases. Using these assumptions the overall estimate of avoidable water related disease is 30 million cases annually in the East European region. The majority of cases occur in the Danube countries and the Russian Federation, which share 17 million cases annually. Two thirds of the predicted disease is general gastro-enteritis.

The benefits of disease reductions have been estimated by an approach which involves the application of benefits transfer to available cost of illness (COI) studies. On the basis of these estimates it was possible to derive best approximations of the unit cost of diseases in terms of the proportion of per capita GDP (gross domestic product). They range from 0.444 (typhoid) to 0.014 % (other gastrointestinal disease). To these estimates of the morbidity impacts, a value for premature mortality effects had to be added. The Value of a Statistical Life (VoSL) estimate used here is three million Euro based on a survey of the Water Research Centre (WRC) of estimates in current use for policy evaluation. As the demand of reduced health risk does not vary proportionately with income and is more intense at lower income levels (Krupnick et al. 1996), an income elasticity of demand of 0.35 has been adopted.

Annual benefits in the East European region of 13 billion Euro were estimated. To a large extent the benefits arise from reductions in disease morbidity (9.6 bn). The highest benefits are seen in the Russian Federation (4.1 bn). The main contributions come from giardiasis and gastro-enteritis. The more severe diseases hardly affect the overall outcome, contributing in sum 0.05 % of the benefits.

Traditional engineering cost models can be used to give an indication of the potential costs of closing the gap between the level of water supply and sanitation in the West and East European region, but scale of required infrastructure as well as local cost aspects have to be taken into consideration. Annual costs of 12.4 billion Euro have been calculated for the total region (35 to 60 Euro per capita) to cover investment, operation and maintenance.

Total benefits exceed total costs, the Russian Federation, the Black Sea and the Danube countries contributing to this net benefit. Furthermore, avoided costs of illness are only one aspect of the health benefits ignoring the pain, suffering and distress associated with the disease. And investing in water supply and sanitation will produce benefits over and above those associated directly with health (altered economic conditions at the macro-level, cleaner environment). Improving water supply and sanitation conditions in the East European region is likely to produce substantial net benefits. These results may be a useful clue for other world regions.

Conclusions

Today, microbial pathogens are regarded to be the most important water related health risk factors. Emerging pathogens as well as outbreaks of well known diseases present a huge health and economic burden. There are still many problems to solve in the developed countries, but for most developing countries waterborne infectious diseases show a dramatic impact.

Lead, nitrate, arsenic, fluorides, pesticides, solvents and aluminium are today world-wide the chemical substances with the largest impact on human health.

Overpopulation and a continuously growing water demand, spatial disproportion of water demand and availability, overexploitation of resources, environmental pollution, climate change and military conflicts around water are the outstanding challenges for water hygiene in the 21st century.

GIS-technology opens new opportunities for the field of water and health. It helps to elucidate locations, time trends, spatial patterns and to model spatial processes. It is currently applied to manage catchment areas, to administer distribution systems as well as to support epidemiological investigations and risk communication. But its application presupposes the generation of basic datasets which is expensive and time consuming. In addition, spatial analysis, interpretation of the outputs, and designing displays for the public needs experience in handling of spatial data and in working at the interface between the natural environment and human activities.

The economic dimension is significant for the field of water and health. The benefits of improving water supply and sanitation conditions in many parts of the world are likely to be immense. Although gross costs are high, such investments are likely to produce substantial net benefits, which will increase where well targeted and locally effective strategies are pursued.

There is a dramatic world-wide situation when concerning water for human consumption and an urgent need of new and holistic strategies:

- Politicians have to recognise and accept the value of hygienic safe water for public health. The UN water assessment points out that looming water crisis must be addressed by political decisions which reallocate water regarding economic and social benefits.
- National policy framework and sectoral strategies should be formulated within a process appropriate for every country.
- Competence has to be increased taking into account the development of human resources, the institutional framework, the legal framework and gender issues.

- The polluter pays principle, peoples participation in decision processes and the enhancement of fiscal incentives have to be established.
- The price of water must reflect its scarcity and environmental costs.
- Green industries should be given higher credit.
- The reduction of fertilisers and pesticides should be encouraged.
- Quality and quality control of sewage and water management must be improved.
- Surveillance of water related illness needs further improvement.
- Sound management practices and appropriate technologies should be applied which effectively maintain facilities and equipment.

Many people, especially children, consuming inadequate water will furthermore suffer from infection and intoxication, if no efforts are made to improve water quality world-wide. Developing countries should gain from the developed countries 19th and 20th century experiences.

References

Clarke KC, McLafferty SL, Tempalski BJ (1996) On epidemiology and geographic information systems: a review and discussion of future directions. Emerging Infectious Diseases 2: 85–92

Dieter HH (1998) Proactive Limit Values for Responsible Management of Chemicals. ESPR – Environ Sci & Poll Res 5: 51–54 (*part I*), 112–116 (*part II*)

Dunn C (1992) GIS and epidemiology. AGI publication number 5/92. Association for Geographic Information, London

Exner M, Böllert F (1991) Hygienische Aspekte der Cholera unter besonderer Berücksichtigung der Epidemie in Südamerika. Bundesgesundheitsblatt 34: 401–414

Gessner BD, Beller M, Middaugh JP, Whitford GM (1994) Acute fluoride poisoning from a public water system. New Engl J Med 330: 95–99

Grimaldo M, Borja-Aburto VH, Ramírez AL, Ponce M, Rosas M, Diaz-Barriga F (1995) Endemic Fluorosis in San Luis Potosi, Mexico. Environ Res 68: 25–30

Hoffmann M (1995) Über Herkunft und Vorkommen toxischer Stoffe im Trinkwasser ukrainischer Großstädte. gwf Wasser/Abwasser 136: 85–90

Kistemann T, Dangendorf F, Koch C, Fischeder R, Exner M (1998) Mikrobielle Belastung von Trinkwassertalsperren-Zuläufen in Abhängigkeit vom Einzugsgebiet. gwf Wasser/Abwasser 139: S17-S22

Krupnick A, Harrison K, Nickell E.,Toman M (1996) The value of health benefits from ambient air quality improvements in Central and Eastern Europe: An exercise in benefits transfer, general environmental value – annual benefits. Environmental and Resource Economics 7: 307–332

MacKenzie WR, Hoxie NJ, Procter ME, Gradus MS, Blair KA, Petersen DE, Katamierczak JJ, Addiss DG, Fox KR, Rose JB, Davis JP (1994) A massive outbreak in Milwaukee of Cryptosporidium infection transmitted through the public water supply. N Engl J Med 331: 161–167

NorthWestWater (1999) The use of Geographical Information systems (GIS) within the NWW. NorthWestWater Water & Health 27: 6–9

Saha et JC, Dikshit AK, Bandyopadhyay (1999) A Review of Arsenic Poisoning and ist Effects on Human Health. Crit Rev Environ Sci Technol 29: 281–313

UN/ECE, WHO/EURO (1999) Protocol on water and health to the 1992 convention on the protection and use of transboundary watercourses and international lakes. Adopted on 17 June 1999 at the Third Ministerial Conference on Environment and Health (www.who.dk/London99/WelcomeE.htm)

WHO (1993) Guidelines for drinking-water quality, 2nd ed./vol. 1: Recommendations; vol. 2: Supporting information. WHO, Geneva

WHO (1995) Concern for Europe's Tomorrow. Wissenschaftliche Verlagsges. Stuttgart mbH

WHO (1999) Geographical Information systems (GIS). Mapping for epidemiological surveillance. Weekly Epidemiological Record 74: 281–285

WHO/EURO, EEA (1999) Water and Health in Europe. European Environment Agency, Copenhagen

WHO/EEA (1999). Water Resources and Health. WHO, Geneva

Wu C, Maurer C, Wang Y, Xue S, and Davis DL (1999) Water pollution and human health in China. Environ Health Perspect 107: 251–256

CHAPTER 15

Urban Thirst for Water and Priorities for Action   15

Surinder Aggarwal*

Water is a fundamental resource on which depend the life support systems and which has to be equitably shared between all those living in a particular river basin (Ayibotele and Falkenmark 1992). It is widely believed that improvement in the supply of water and sanitation can play a major role in improving the lives of the poor in developing countries. The industry sector in the West has realised the importance of depleting fresh water resources for their survival and sustainability and has given a wakeup call to the sector. The outcome of this realisation is that UNEP and the World Business Council for Sustainable Development (SD) have come out with a report "industry, freshwater and SD". Apparently, the world's thirst for water is likely to become one of the most pressing resource issues of the 21st century, if the present trends of consumption and management continue (World Resources Institute et al. 1998). During the preparation of Global Environment Outlook-2000 (*GEO-2000*), a global survey on emerging environmental issues was conducted by ICSU's Scientific Committee on Problems of the Environment (SCOPE) as part of the GEO programme. Climate change was the most cited issue in the SCOPE survey although, taken together, water scarcity and pollution ranked higher (UNEP 1999).

Since the commitments of the **International Drinking Water Supply and Sanitation Decade** of the 1980s have not been fully met both for rural and urban settlements, this calls for the highest priority to be given to the serious freshwater problems facing many regions, especially in the major cities of the developing world (Christmas and Rooy 1991, McGary 1991).

Growing urbanization trends in the developing countries have shifted the water demand from agriculture sector to the industries and domestic sectors and are a pointer to the rising water conflicts in the 21st century (UNEP 1999). Intra-urban inequality in water supplies, favouring the rich areas, is going to be another area of social unrest and political conflict within the large cities. The poor, in particular, find them marginalised and perhaps pay more in proportion to their income for their water needs. The heightened water withdrawals from the aquifers and discharge of untreated waste water into the fresh water bodies, particularly from the megacities of the world, has raised concern over environmental degradation of land subsidence and water pollution (ecological footprints) within the cities and their water resource regions (Munasinghe 1992, Rees 1992, UNESCAP 1993).

* e-mail: sagarwal@giasdl01.vsnl.net.in

Poor water management practices including water pricing, subsidies, cost recovery, unaccounted for water and lack of integrated water management efforts is another major issue that needs to be addressed for understanding the underlying causes of urban water thirst at least in the developing countries. The developed countries of course do not suffer much from these availability, equity and management issues but are concerned about the deteriorating water quality and other ecological footprints. These concerns along with possible priorities for action were presented in the Plenary Sessions in general and in the Break-Out Group 2 in particular and are addressed in this paper broadly to mitigate the future urban water stress.

Current fresh water situation

The myth that water is in abundance and inexhaustible and should be available either freely or at little cost needs to be demystified if we examine the trends of present water consumption and the availability of fresh water resources. While water covers nearly 72 % of planet, only less than 1 % of the world's fresh water (or about 0.007 % of all water on earth) is available for human consumption (WHO 1999). Much of this fresh water is not accessible for exploitation costs and ecological considerations. Further, freshwater resources are unevenly distributed. Arid and semiarid zones, holding bulk of the global population and 40 % of the landmass, have only 2 % of the global run off (World Resources Institute et al. 1996, WHO 1999).

At the same time if we examine the water demand in various sectors of the economy, domestic needs and ecosystems needs, it is apparent that we are heading towards water crisis situation in the twenty first century. Some of the astonishing water statistics tell us the present grim situation and the crisis the world is going to face in the next fifty years. Per capita water consumption rises twice as fast as world population (Speets 1999). Global freshwater consumption rose sixfold between 1900 and 1995 – more than double the rate of population growth (UNEP 1999). In 1995 only 8 % of the world population had insufficient water and it is projected to rise to 42 % by 2050. 400 million people face water shortage today and this number will rise to 4 billion by 2050. About one-third of world population at present lives in countries with moderate to high water stress (where water consumption is more than 10 % of the renewable freshwater supply (UNEP 1999). Not only the humans but also the various ecosystems, essential for the survival of the planet earth, are also threatened by excessive water use. According to a recent comprehensive assessment of the fresh water resources of the world by the World Meteorological Organisation (WMO), the ecosystem would be starved of substantial replenishment in the next 50 years. Apparently, with a projected increase of 50 percent in population in the next 50 years and the expected increase in demand as a result of economic growth and changes in life styles, the future does not look very bright, unless there is a very strong planning and management of water (Centre for Science and Environment 1998). The major driving forces behind the present and future water crisis according to Speets (1999) include overexploitation and pollution of fresh waters, population growth and hyper-urbanization.

Urbanisation and shifts in water demand

Urbanisation trends: developed and developing countries

The world is in the midst of a massive urban transition. Within the next decade, more than half of the world's population, an estimated 3.3 billion will be living in urban areas – a change with vast implications both for human well-being and for the environment (UN 1995). Population studies also indicate that by 2025 the total urban population is projected to double to more than 5,000 million people (two-thirds of the world's population) out of which 90 % will be in developing countries (UN 1997). The most rapid change is occurring in the developing world, where urban population are growing at 3.5 % per year, as opposed to less than 1 % in the most developed regions (UN 1995). One distinguishing feature of this rapid urban growth has been the unprecedented growth in number and size of the big cities. The number of megacities (population greater than 8 million) has swelled from 2 in 1950 to 23 in 1995, 17 in the developing world. By 2015, the number is projected to grow to 36; 23 of these megacities will be located in Asia (UN 1997). With this kind of urbanisation trends, especially in the developing countries, the implications for a resource like water have serious environmental, hydrological and social implications since cities in general consume more resources per capita than the rural areas. The rising share of urban population in the shanty settlements of the developing countries would further produce water allocation related conflicts within the big cities.

Sectoral shift in water demand and conflicts

The total available blue water, or renewable water resources, is about 40,000 cubic kilometres (km^3 or billion m^3) a year. Of this, some 3,500 cubic kilometres, roughly 10 % are "withdrawn" (diverted or pumped) for various human uses. Of the water withdrawn, more than 2,000 cubic kilometers are "consumed", with the remainder returned, usually with significant reductions in quality (Cosgrove and Rijsberman 1999).

Total water demand is primarily the sum of the demands in the domestic, industrial and agricultural sectors. Urban demands in both the domestic and industrial sectors are largely dependent on the magnitude of a population, changing life styles and level of economic development (Appan 1999). The phenomenal growth in urban population has affected the share of fresh water demand (withdrawal and consumption) in various sectors of economy and other uses (see table 1).

Withdrawals for irrigation in 1990 were in the order of 70 % of the total withdrawn, those for municipal/residential use about 10 % and those for industry 20 % (see table 1). Irrigation "consumes" a large share of the water it withdraws.

Table 1. Drying up: estimated and projected global water use (cubic kilometers per year)

Sector	1950	%	1990	%	2025	%
Agricultural use						
Withdrawal	1,124	82.3	2,412	67.3	3,162	60.9
Consumption	856		1,907		2,377	
Industrial use						
Withdrawal	182	13.3	681	19.0	1,106	21.3
Consumption	14		73		146	
Municipal use						
Withdrawal	53	3.9	321	9.1	645	12.4
Consumption	14		53		81	
Reservoirs (evaporation)	6	0.5	164	4.6	275	5.4
Total						
Withdrawal	1,365	100	3,580	100	5,187	
Consumption	894		2,196		2,879	

Source: Shiklomanov (1997).

A very large share of water withdrawn for municipal or residential use (up to 90 %) is returned as wastewater, but often in such a degraded state that major cleanups are required before it can be re-used. Industry consumes a little over 10 % of the water it withdraws, heavily polluting a small fraction that is returned. From the above table we find that the share of agriculture sector in total water withdrawal declined to 67.3 % in 1990 as compared to 82.3 % in 1950, whereas for the industrial and domestic sectors combined (urban sector demand) it increased from 17.2 % to 28.1 % for the corresponding period. This trend is projected to continue in the next 25 years. Industrial and domestic water demand in the urban areas is expected to double by 2025 if current growth trends persist. Of late, the ecosystem is also being viewed as a user of water. A recent report on *comprehensive assessment of the fresh water resources of the world* prepared by WMO mentions the need of water to be left in the rivers to maintain healthy ecosystems and if we go by the water use projections the ecosystems would be starved of substantial replenishment in the next 50 years (Centre for Science and Environment 1998).

These shifts in water demand, particularly in favour of urban sector, are causing and expected to exacerbate water conflicts between the rural and urban sectors and as well as between the industrial and domestic sectors within the urban settings. For example, Western United States has to face fierce conflicts between urban-industrial users and the agricultural sector; the domestic and hydropower generation in Manila (Philippines) is creating water shortages in Central Lurzon; increasing water demands for the city of Hyderabad in India are having an adverse impact on irrigation of its hinterland are some of the inconveniences present in the distribution of water among rural and urban sectors (Biswas 1996).

Major water issues in urban sector

Water Consumption

Despite the economic and other social opportunities which cities provide, they are heavy consumers of resources in concentrated space resulting into serious environmental hazards and degradation, if not managed properly. Resource consumption, like water, varies with the level of income and wealth. As with most other forms of consumption, that of water is much higher in cities of higher income countries. For example an urban dweller in New York consumes approximately four times of water than does a resident of Bombay (see fig. 1). Typically people in cities of developed countries use 272 litres of water per day, while the average in Africa is 53 litres per day. At the city level inequities exist between various rich and poor areas with respect to duration and amount of water supply. In case of Mexico city although 94 % of the inhabitants are served with a water connection, low income areas like Chalco use between 20 and 80 litres/person/day while consumption in selected rich areas ranges between 1,000 to 2,000 litres/person/day (Lankao 1999). And, on average, cities in North America and the former Soviet Union use twice as much water per person as do Western European cities, and seven times the rate of African cities (Urban Age 1999b).

Efforts are needed to regulate the use of water by pricing instruments, public awareness about water as a scarce resource, reduction of subsidy, community participation in water management, reuse and recycling the waste water and application of improved technology to reduce the water consumption in various sectors of the economy including domestic sector.

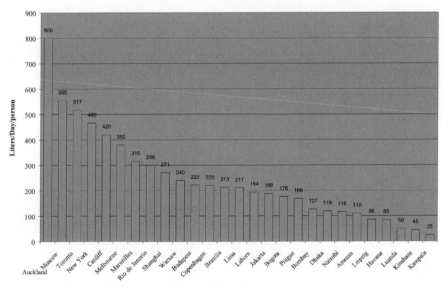

Fig. 1. Water Use in Major Cities of the World
Source: World Resources Institute et al. (1998) pp 278–279.

BOX 1

Water supply in Indonesia's Megacities.
In the city of Jakarta, the water supply and disposal systems were once designed for half a million people. In 1985, the city had a population of 7.7 million and a projected growth to 17 million by the year 2000.The city currently suffers continuous water shortage and less than a quarter of the population have direct access to water supply systems. Over-exploitation of groundwater increases saltwater intrusion in the aquifers and further accentuates the water supply problems.
The water level in what was previously an artesian aquifer is now generally below sea level; locally it is down to 30m below. Saltwater intrusion has ruined this as a source of drinking water.
Source: Sundblad (1999).

Access to Water

Access to adequate clean water at home or within a reasonable walking distance (200 meters) by all residents can be a good indicator of equity and urban water management. Unfortunately this is not happening at least in the cities of the developing countries and has become a matter of serious health concern and of a basic human right. Availability of potable water in urban areas increases rapidly with income (see fig. 2). In Luanda close to one-third urban residents have access

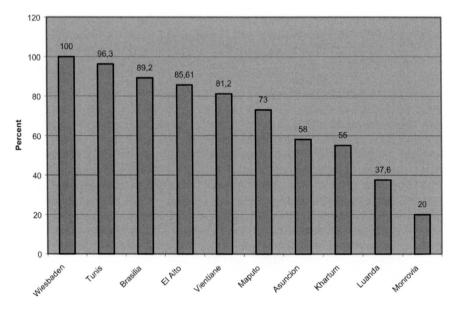

Fig. 2. Access to potable water within 200 meters
Source: UNCHS (1996).

to clean water whereas Wiesbaden residents have 100 % access. At least 30 % of the households in the lowest income countries do not have access to clean water within 200 meters distance, while almost everyone living in cities in developed countries has such access and every household is connected with municipal water system (Urban Age 1999b). Women and children in the developing countries suffer more due to inaccessibility of water.

In the eighties during the Water and Sanitation Decade more than 1.6 billion people gained access to water of reasonable quality, and also sanitation coverage increased substantially (Speets 1999). Based on WHO data, Speets indicates that despite the advances made in global water supply coverage during and since the Water Decade, at least 170 million people in urban and 850 million in rural areas still lack access to a source of safe, potable water, and more than 2 billion people lack adequate sanitation. With some 900 million people affected by diarrhoea each year, and an equal number suffering from diseases by roundworms, unsafe water ranks at the top of the world's health problems. While discussing the equity question in water in South Africa, Maharaj and Pillay (1999) share the agony of many black households whose access to adequate safe water is through a shared standpipe some distance away. The White Paper on water supply and sanitation also emphasised the relationship between poverty, race and access to water in South Africa.

Pricing, cost recovery and subsidy issues

Water has been perceived in the past an inexhaustible resource and available in abundance due to its recycling nature. Being a basic need for biological survival and to maintain health, and also an essential input to agricultural and industrial growth it has been usually considered a social good and hence underpriced and heavily subsidised. According to a World Bank study, consumers pay only 35 % of the cost of supplying water in many developing countries. For example the marginal cost of supplying water to Mexico city is estimated at about $ 1.00 per cubic meter, but only $ 0.10 per cubic meter is collected from water users (Lankao 1999). With growing population pressure and high urban growth rate in the developing countries, the demand for water has accelerated and it will be difficult to meet the rising demand without increasing the water price. More recently there is a growing recognition that unless water is considered an economic good along with its social function its overexploitation and wasteful use will continue (UNESCAP 1993).

On matters of removal or lowering of water subsidy to urban consumers and cross subsidy to the poor there is a debate. Cross-subsidy or subsidy has been normally suggested a viable and socially just solution to serve the low-income consumers. Maharaj and Pillay (1999) like many other experts from the developing countries favour continuation of subsidy considering unaffordability aspect of the poor. However, such subsidy systems are not necessarily socially benign. On the contrary, there is evidence to suggest that the structure of subsidies is often highly regressive, with substantial benefits going to middle and higher income groups. Rather the poor end up purchasing the water from private vendors, often paying prices well in access of those associated with the supply from the public network (Foster 1998). Low-priced water encourages excessive consumption by those connected to the supply system. This limits utilities' coverage

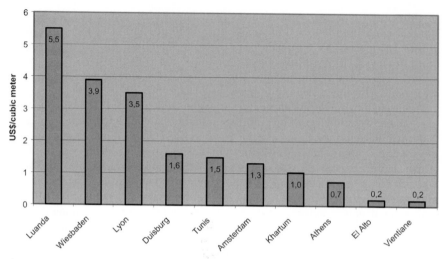

Fig. 3. Median 1993 Price of Water
Source: UNCHS (1996).

and, as a result, the poor are left to purchase higher priced water from vendors. Indeed the poor may pay as much as 30 % of their income for water, while the well-to-do pay less than 2 % (Okum1988). The median price of water in the cities often varies with their location, level of development and the subsidy component. Prices do not strictly vary with the level of economic development as is evident from figure 3. Vast literature on pricing of water indicates that both intercity and intracity differential exist. The inequality is more pronounced in the cities of the developing countries where the poor, not hooked to the city water system, end up paying much higher than the rich (Box 2).

BOX 2

It's expensive to be poor

In Port-au-Prince, Haiti, a comprehensive survey showed that households connected to the water system typically paid around $1.00 per cubic metre, while unconnected customers forced to purchase water from mobile vendors paid from $5.50 to a staggering $16.50 per cubic metre.

Urban residents in the United States typically pay $0.40–0.80 per cubic metre for municipal water of excellent quality. Residents in Jakarta, Indonesia, purchase water for between $0.09–0.50 per cubic meter from the municipal water company, $1.80 from tanker trucks and $1.50–2.50 from private vendors – as much as 50 times than residents connected to the city system.

In Lima, Peru a poor family on the edge of the city pays a vendor roughly $3.00 per cubic metre, 20 times the price paid by a family connected to the city system.

Source: Water Supply and Sanitation Collaborative Council (1999).

Although per capita consumption in the cities of the developed countries is high, subsidy on water is relatively much lower and thus most costs are recovered through pricing mechanisms. This is almost absent in the developing countries for political reasons and high prevalence of poverty. The policy makers in the developing countries has to act soon to develop a realistic price structure, user-pay principle or targeted subsidies approach for the poor to overcome the looming water stress situations in the coming years. Many studies indicate rather confirm that the poor are willing to pay provided they get a title of the land and assured water supply (World Resources Intsitute et al. 1996, Centre for Science and Environment 1998)

Pricing and subsidy, therefore, are among the most crucial instruments in disciplining the consumers towards conservation of fresh water and other ecological needs. This will remain one of the greatest challenges for many municipal governments in the developing countries to discipline their consumers who have enjoyed the water subsidy regime.

Urban water use and environmental impacts

Cities affect more than the areas they occupy. Their "ecological footprints" can be enormous because of their huge demands for food, energy, water and other resources from the surrounding hinterland (Rees 1992). In order to save itself from the diminishing trend of carrying capacity, the city creates a stress on its neighbouring areas to accommodate its thirst, but jeopardise the basic wants of these surrounding areas. Disposal of their wastes into surrounding soil, air and water are becoming equally alarming and hazardous for the local and regional natural ecosystems.

In recent years the phenomenal increase in demand for fresh water in the cities of both developing and the developed countries has caused ecological alarm with respect to over-extraction, excessive use and pollution of both surface and ground water. Over exraction of water without commensurate replenishment has disturbed the local hydrological cycles and perforce the cities resort to import water from distant river basins. Cities like Dakar, Los Angeles, Mexico have to supplement their ground water supplies by bringing water from distant river systems and reservoirs as far as 200 km (Urban Age 1999a, White 1992). The steep decline in water tables caused by rapid ground withdrawals has led to land subsidence and intrusion of salt water into the aquifers in many cities. In Thailand, the rapid lowering of the water table in Bangkok has resulted in the contamination of the aquifers in the surrounding regions with salt water intrusion from the nearby ocean. Identical problems have also occurred in Jakarta, Madras and Manila (UNESCAP 1995). In Europe, the need to find clean water supplies has led to the extraction of deep ground water around cities; in Germany, this has caused irreversible changes of heath land around cities. Further, the over-abstraction of water from both ground and surface sources has reduced the replenishment of the aquifers leading to the disruption of local hydrological cycles, at least around the major megacities. This has a serious impact on the flow of rivers, especially during dry periods, so vital for the aquatic ecosystems. In the case of the river Yamuna in India, overextraction of freshwater has denied the river of the minimum flow it requires to actually "cleanse" itself (Centre for Science and Environment 1998).

Pollution is another major factor that is reducing water quality and thereby the water availability of clean water. Discharge of wastewater, with little or no treatment, has polluted the water bodies and restricted their direct use for different economic activities and human use. Wastes discharged into the water bodies have outstripped nature's ability to break down pollutants into less harmful elements. Seoul faces a degrading water supply due to waste water inflows from residences, industries and livestock farms in the watershed of the Padlang Reservoir (UNESCAP 1990).

Physical and commercial leakages

Water disappears from city systems around the world through a variety of physical and commercial leakages including theft, inadequate metering and billing, inefficient collection procedures and poor maintenance of the pipes. Physical leakages in the range of 5–10 % are common in the cities of the developed countries, whereas the cities in the developing countries suffer high losses from both commercial and physical leakages. In the Third World cities, as much as 60 % is lost through leaky pipes and theft (O'Meara 1999). In Hanoi, for example, 63 % of the drinking water is forfeited to leaks or illegal tapping, whereas Singapore, where pipes are better maintained, loses only 6 % (see fig. 4). We also observe from figure 4 that water availability duration is usually much higher in cities where unaccounted for water loss is low. Water leakages are also linked with water quality. It is virtually impossible to maintain safe water in a distribution system with intermittent pressure or less than 24 hours supply with severe leakage losses. How do governments justify expenditures of funds for treatment and supply of water when up to 60 % is lost.

If we consider these losses seriously and local governments take measures to reduces these loses within an acceptable level say 10 %, there is less need to augment water supplies through exploiting additional resources and these water savings can comfortably meet large part of the additional demand created by the rising population or other reasons. Cities that have aggressively attacked the problem report considerable benefits. Singapore for example, brought the percentage of the lost water due to commercial and physical leaks down from 10.6 % to 6 % in six years by controlling tampering, installing meters on all outlets and ensure that the entire water distribution system is swept for leaks once a year and devices are replaced every four to six years (Urban Age 1999a).

Paper summaries

Presented below are the summaries of the presentations at the Break-Out Group on Urban Thirst and Water Conflicts. The intention is to present the studies in a manner that we understand the contemporary issues prevailing in the urban water demand in different geographical settings and the lessons that can be learnt for better management of water in the urban areas.

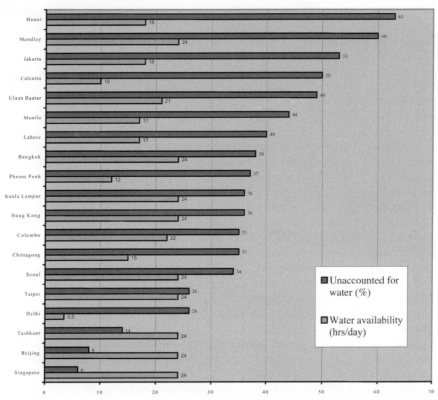

Fig. 4. Availability of Water and Unaccounted for Water in Major Cities in South and South-East Asia
Source: Asian Development Bank (1997).

Patricia Romero Lankao: Water in Mexico City: Trends and Challenges

Getting drinking water to the inhabitants of one of the world's largest cities poses almost every imaginable problem, and making that water safe to drink only complicates matters. In her paper on **Water in Mexico City: Trends and Challenges**, Patricia Romero Lankao addresses the historical development of water management schemes, disturbed hydrological cycle, environmental impacts (ecological footprints) of water systems development in the Mexico city metropolitan region and beyond, equity questions of distribution and water management tools and the challenges ahead.

Abundant ground and surface water resources and floods characterised Mexico City's water system in the past. Historically, indigenous inhabitants and early Spanish governments developed efficient schemes of managing water by combination of public and private partners in using and disposing water of Mexico City while taking advantage of small population base and limited water demand. In later years (1876–1990s), growing deforestation, massive population growth

(20 million at present), economic dynamics accompanied with urban sprawl overexploited the surface water resources such as the existing lakes and river basins through a network of dikes, canals and drains. At present nearly 70 % of the total water demand in the metropolitan region is met by under ground aquifers and the rest by the Lerma and Cutzamala river basins, 120 kilometres away and pumped 2,500 meters above sea level to reach the Valley of Mexico. Despite the feats of engineering, the city can not meet the needs of an ever-growing population. Apparently, the federal government controlled supply side strategies supported with an extensive and well-developed network of water infrastructure are not working. The cost recovery is poor, commercial and physical leakages are more than 25 %, and water has become polluted and new sources becoming scarce. The conventional engineering solutions are expensive and beyond the reach of the poor.

According to the author, heavy deforestation, sprawling built-up area, the phenomenal water demand of 60 cubic meters per second has disturbed the local hydrological cycle as ground aquifers are not being replenished due to lack of infiltration and heavy runoff. A combination of increasing ground water extraction and artificial diversions to drain the valley resulted in the drying of the local and regional lakes system and of many springs. Overexploitation of aquifers at the rate ranging between 16.9 and 21.1 cubic metres per second has resulted in average annual declines in ground water levels from 1.1 meters to 1.5 meters per year in the heavily pumped areas of the Basin. Heavy extraction has provoked land subsidence up to 46 centimetres per year in some areas and it has caused dislocation of water infrastructure and buildings and accelerated the downward movement of surface contaminants.

The problem of water quality is very much related to the problem of water scarcity. Discharge of untreated waste water is polluting the soils, water bodies and vegetation and has thus affected the regional cropping pattern. To protect public health related to regional pollution, authorities have promoted crop restrictions of tomato, carrot, lettuce and other leafy vegetables, rather than wastewater treatment. According to the national statistical institute even though 100 % of city's tap water is treated, still 585 persons per 100,000 residents died of infectious intestinal illness in 1995 (Urban Age 1999a). Even though nearly 94 % of the population of Mexico city is accessible to city water supply, the inequity in service level (quantity and quality aspects) prevails to the detriment of the poor and residents of disadvantaged peri-urban settlements.

To counter these impacts and improve water management, National Water Law and its regulation have been modified to promote private water rights, privatisation of management of water supply and wastewater services and the requirement to conduct cost-benefit analysis in the application of regulatory standards. Demand management tools such as retrofitting of plumbing equipment, increases in water tariffs have been invoked, and are aimed at controlling water consumption and modifying behaviour of water users. Still the author believes that more needs to be done in the following aspects:

- Reduce environmental and social impacts of water systems.
- Develop a comprehensive compensation or assistance mechanism to households affected by the development, construction and operation of water systems.

- Frame policy instruments towards reusing and recycling of water and changing social patterns of water use.
- Promote public education campaigns towards water conservation and efficient use of water.

The major lesson learnt from this study suggests that supply side conventional engineering solutions are expensive, anti poor and difficult to sustain. Community participation and low cost technologies, not suggested by the author, are possible viable solutions for water and sanitation problems of the megacities like Mexico.

A. Appan: Quenching Urban Thirst in the New Millennium: Singapore Case Study

In a different geographical setting the study on Singapore presents an interesting approach to quench thirst of a giant and land constraint metropolis. Appan is of the view that the major water demands to be met in the next millennium will be from urban populations in fast-growing megacities in the developing countries. The category of people most affected in the megacities will be the urban poor who will have limited access to safe water and will also have to pay more.

The major contribution of the paper remained to share experience of Singapore, a small country that has limited water resources but has an ever-growing urban thirst. Much fresh water demand (40–50%) of Singapore is met with import from Malaysia, the rest is managed within the country by adopting innovative water harvesting, conservation and recycling methodologies. Due to competing demands for the limited land area, all possible measures are being taken to maximise the use of the land in relation to the annual rainfall of about 2400 mm. Successful projects have been implemented in Singapore where urban water has been successfully tapped. These schemes include systems that cover large urban catchments (greater than 1000 hectares) and also smaller rainwater catchments systems (less than 1000 hectares) wherein runoff is collected from roofs of high rise buildings, factory lots, airports, public buildings etc. The urban storm water collection system in Singapore is a large-scale project which entails the use of large tracts of highly urbanised area and the water abstracted is comparable to that in conventional catchments.

The main reason for such a success in urban water projects is the rigorous measures taken to keep the catchments clean during the planning and subsequent periods of operation. This exemplifies the feasibility of the development and use of urban catchments for the abstraction of drinking water. Most of the rainwater catchment systems have shown that smaller catchments can be fruitfully utilised to develop water supply schemes catering for non-potable uses. Such schemes help to fully utilise urban runoff that will otherwise go waste and to reduce potable water consumption. Simple systems for roofwater collection are also available which have been implemented in developing countries like Indonesia and Thailand. Such systems will be low priced, self-sustaining and most suited for the urban poor. On the whole, it is conceivable that the development of urban catchments for abstraction of water is both technologically feasible and economically viable as successfully demonstrated in Singapore.

It can be learnt from Singapore experience that if a proper approach is taken towards the salient factors of designs which largely are dependent on excellent inter-departmental co-operation to keep catchments clean, such successful low-cost technology water supply systems can be established in the potential urban areas of megacities of the future.

Brij Maharaj and Salosh Pillay: Urban Water Issues and Policy in the New South Africa: A Critical Analysis

Maharaj and Pillay assess urban water issues, policy and strategies in South Africa in the context to pre- and post-apartheid situations. According to the authors, a key issue to be addressed with regard to service provision in the new South Africa is equity. The most serious problems are in the black townships "with high population densities and polluted waters sources". The authors maintain that lack of access to basic water supply and sanitation does not exist in isolation from other poverty-related issues. The White Paper on Water Supply and Sanitation in South Africa emphasised the relationship between poverty, race and access to water. For women, the drawing and carrying of water forms a substantial part of their daily effort to survive.

- South Africa's apartheid history has produced a land of inequalities and contradictions, and this is especially apparent in the delivery of basic services like water and sanitation. Pre-apartheid regime characterised in inequality in allocation and distribution of water based on race. Under apartheid water legislation was "written largely in the interests of commercial agriculture and industry, which necessarily excluded a large sector of the black population".
- In the post-apartheid South Africa, providing an adequate level of water and sanitation service under the Reconstruction and Development Programme (RDP) has been given an important priority and is recognised as a basic human right. However, the reality of historical backlogs, resource constraints and widespread poverty means that millions do not have access to basic water and sanitation services even now. Many black household's only access to adequate water is through a standpipe some distance away.
- With the adoption of the neo-liberal structural adjustment strategy (GEAR) in South Africa, user-pay principle, the principle of cost recovery, public-private-partnership, privatisation and reduction of subsidy are considered more pragmatic and sustainable instruments to improve the level of basic services, including water. Based on international experience, the authors, however, believe that the urban poor have suffered most from such restructuring and adjustment strategies and hence the subsidies should not be withdrawn and should be targeted towards the poor. On the contrary, the governmental view is that subsidy is counter productive for the poor since the non-poor enjoy more advantage of the subsidy under the prevailing power relations.
- Further, the authors are of the view that the problems of urban water availability may be further compounded since South Africa falls at present in periodic or regular water stress category and heading towards chronic water scarcity. In addition to its own supplies, South Africa also draws on water supplies from

its neighbours and other studies warn that this has the potential for geopolitical conflict.

Jan Speets: Poverty, Health and Water Resource Management in Urban Thirst and Conflicts

The paper focuses on three essential issues with regard to existing and forthcoming problems for drinking water supply in cities: poverty, health and water resources management. The paper brings out a number of related issues, such as water economy, inequitable services, decentralisation of water management etc.

Linking health, poverty and water

- Health, poverty and water resources management is interlinked. Ill health often caused by lack of safe water brings poverty.
- Present water supply systems put the poor in a very disadvantaged economic position: for the same quantity of water of less quality, they pay more than those having the prerogative of being connected to the municipal piped distribution system, or having their own private supply.
- It is highlighted that corruption and political interference constitute a major cause for not receiving adequate and appropriate basic services for low income communities.

Root causes of Water crisis in South and South-East Asia

- System of water rights to landowners
- Uncontrolled borewell construction
- Community is not in control of their water resources
- Pollution of freshwater sources
- Inadequate:
  - Water conservation practices
  - Efficiency in water use
  - (Waste) Water re-use
  - Ground water re-charge

Strategies to address water crisis:

- Decentralise management of water resources to the community
- Enhancing community awareness and management of fresh water resources
- Effective ground and surface water legislation and regulations to be implemented by communities and local institutions
- Water quality to be the central issue in designing and implementing programs
- External support agencies to support fresh water resource management
- Define basic services levels and technology options commensurate with the users
  WHO's Healthy Cities program is given as an example on anticipating and resolving urban problems and conflicts, by placing health and environment issues at the heart of sound economic urban development

Jose A. Diaz: Water Quality Automatic Information System in Tajo Basin (Spain)

Monitoring water quality and its surveillance are major environmental health concerns in most industrialised countries. Heavy industrialisation, thermal plants and intensive use of chemical fertilisers and pesticides in agriculture are producing threat to the water bodies that are vital from ecological perspective and also from a human consumption viewpoint. Even though water treatment facilities exist to basically remove most biological organism from domestic waste water and chemical effluents from most industrial activities, still excessive chemical load is threatening the water ecosystems in many industrial economies. Therefore, constant monitoring of industrial and domestic wastewater has become an essential element in supply of safe drinking water for the health of individuals and various water ecosystems. In addition, water monitoring is considered necessary in Spain for the following considerations:

- **To promote sustainable water use based on a long term protection of available water resources**
- **To prevent further deterioration and protect and enhance the aquatic ecosystem**
- **To mitigate the effect of floods and droughts**
- **To provide sufficient supply of quality water**
- **To maintain water quality and quantity using new technology**

With this background, the author elaborated upon the continuos survey mechanism established in Spain for monitoring water quality. In Spain the existence of a network of 250 monitoring stations assists the automatic water quality control. The monitoring stations send information via satellite every 15 minute. Use of leading technologies, mainly in analytical and telecommunication aspects, and constant research inputs (BOD, microorganism in particular) add to the quality of monitoring in Spain. River basin is considered the natural unit of water monitoring and surveillance and the states involve co-operate in its sustainable use.

The specific case study on Tajo River Basin was presented in detail to illustrate the monitoring system. The participants discussed that such monitoring systems are quite expensive for the developing countries. Also, even if water is monitored for its quality using these technologies, lack of compliance restricts the use and relevance of such expensive surveillance techniques. The developing countries, therefore, require less expensive alternate technologies to monitor their water quality.

We present below the most critical issues, methodological approaches for sustainable water management, contemporary research challenges and desirable actions that have emerged from the panel discussions, views of the participants and the specific studies presented in the Break-Out Group on Urban Thirst and Water conflicts.

Emerging critical issues and future trends

- Water pricing and subsidy debate towards cost recovery and social obligation.
- Urban poor are being marginalised in access to fresh water and normally pay more than the rich in comparison to their income levels.
- Many rapidly urbanising countries will enter moderate to high water stress categories in the next 25 years.
- Water scarcity at the source caused by disruption of the local hydrological cycle.
- Cities are facing serious shortages of safe water primarily as a result of
  - Overexploitation and pollution of fresh water resources
  - Community not in charge of their water resources
  - Lack of awareness to water conservation
  - Commercial and physical leakages
  - Water as an under-priced service
- Water Use and City Ecological Foot-Prints
  - Depletion of ground water through over-pumping
  - Downstream water pollution and health consequences
  - Withdrawal induced land subsidence
  - Contamination of the aquifers
  - Disruption of the local hydrological cycle
  - Nature's inability to break down pollutants into less harmful elements
- Shift in Water Use and Conflicts
  - Share of demand for water is declining for agriculture and rising for the industry, commercial and domestic sectors and shall become responsible for most water allocation conflicts
  - Developed Economies: rise in chemical pollution and damage to water based ecosystems and human health
  - Developing Economies: rise in both biological and chemical pollution (double-edged sword syndrome)

Methodological approaches for sustainable water management

- **Decentralise** management of water resources to the community.
- **Watershed management** of the river basins for most efficient use of water resources to meet the growing urban water demand.
- Encourage **Public-Private-Partnership** for better management of water systems (pricing and cost recovery) and to overcome financial constraints in new water projects. A Build-Own-Operate-Transfer (BOOT) approach, most popular in other infrastructure projects, can be tried in water sector projects to make them financially and managerially viable.
- Identify **low-cost technology** indigenous solutions towards efficient water use and wastewater disposal and reuse.
- Make full use of **economic instruments** – marginal cost pricing for example – that treat water and land as scarce resource (natural capital).

- **Conservation measures** towards collection, storage and use of rainwater and urban storm water runoff through improved and low-cost harvesting system designs as promoted in Singapore, Thailand and the Philippines.
- Water conservation through repairing leaky pipes.
- Mega cities water **demand forecasting** due to the rising slum population, growth in various sectors of city economy and rising living standards.
- Improve **energy efficiencies** in supplying water.

BOX 3

Water Conservation in Metropolitan Boston

Metropolitan Boston provides an example of successful water conservation. Since 1987, the Massachusetts Water Authority has managed to avoid diverting two large rivers to augment supply, as engineers had initially prescribed. For a third to half the cost of conversions, the government has reduced total water demand by 24% by repairing leaky pipes, installing water-saving fixtures, and educating everyone from school children to plant managers on water saving measures.

Source: O'Meara (1999).

Global agenda for research

- Operational research in water supply and environment sanitation in developing countries faced with choices as to where and how many resources can be allocated to different sectors of economy.
- Establish relationship between minimum service and consumption levels.
- Whether there is competition between domestic and industrial/commercial use of water, and how this influences the access of vulnerable groups.
- How to develop integrated water systems?
- Pollution of water sources and disease burden down the stream (city ecological footprints).
- Structural damage to water ecosystems resulting from excessive ground withdrawals and dumping urban waste water in water bodies (city ecological footprints).
- The water and sanitation needs of the elderly, the poor and new migrants who are very vulnerable.
- The amount of time that women in cities save when water is provided, and how they apportion that time.
- Low cost water quality monitoring and water systems surveillance.
- Need to examine whether existing infrastructure levels are adequate to ensure that there are sufficient levels of consumption to protect public health.
- Incorporate new techniques of data collection and analysis such as space analysis of data through Geographical Information Systems (GIS) application.
- Improve cost/benefit analysis techniques in comparing water supply and sanitation interventions with other measures (such as vaccination).

Desirable actions

- Encourage political commitment to consider water as social and **economic good**.
- Ensure **equity** in access to water for all residents.
- Improve **management** of water resources and increase the availability of water-for urban use.
- Reduce **human health hazards** due to contaminated water.
- Improve the **knowledge base and skills** of administrators and decision-makers in developing countries and countries in transition towards integrated water management.
- Increase **awareness** of needs for efficient water use in urban areas towards improved health.
- Compile available **technology options** and **sound practices** for efficient water use.
  Profile **case studies** where more efficient water use and water resource management practices have been applied.
- Increase **community participation** in awareness generation on reduced consumption, commercial and infrastructure leakage and pricing of water
- **Community empowerment** and involvement in water management at micro level.
- A sound **tariff structure and policy** with clear guidelines for implementation.
- Follow **Dublin principles for managing water:**
  - Fresh water is a finite and vulnerable resource, essential to sustain life, development and the environment.
  - Water development and management should be based on a participatory approach, involving users, planners and policymakers at all levels.
  - Women play a central part in providing, managing and safeguarding of water.
  - Water has an economic value in all its competing uses and should be recognised as an economic good.

References

Appan A (1999) Quenching urban thirst in the next millenium. Unpublished paper prepared for International Conference on Understanding the Earth System, 24–26 November 1999, Bonn, Germany

Asian Development Bank (1997) Second water utility data book, Asia and Pacific Region 1997

Ayibotele NB, Falkenmark M (1992) Freshwater Resources. In Dooge JCI et al. (eds) An agenda of science for environment and development into the 21st century. Cambridge University Press, Cambridge

Biswas AK (1996) Water Development and Environment, in Environment Planning and Management. Mc Graw Hill, New York

Centre for Science and Environment (1998) Special report on water: how sustainable a partner? Down to Earth 7 (1): 24–28

Christmas J, de Rooy C (1991) The decade and beyond: at a glance. Water International 16 (3): 127–134

Cosgrove WJ, Rijsberman FR (1999) World water vision: making water everybody's business. World commission on water for the 21st century. Draft Report of the Commission, Version of 14 November 1999

Diaz Jose A (1999) Water quality automatic information system in Tajo Basin. Unpublished paper prepared for International Conference on Understanding the Earth System, 24–26 November 1999, Bonn, Germany

Foster V (1998) Water and sanitation programme: considerations for regulating water services while reinforcing social interests. UNDP, World Bank

Lankao PR (1999) Water in Mexico city: trends and challenges. Unpublished paper prepared for International Conference on Understanding the Earth System, 24–26 November 1999, Bonn, Germany

Maharaj B, Pillay S (1999) Urban water issues and policy in New South Africa: a critical analysis. Unpublished paper prepared for International Conference on Understanding the Earth System, 24–26 November 1999, Bonn, Germany

McGary MG (1991) Water supply and sanitation in the 1990s. Water International 16 (3): 153–160

Munasinghe M (1992) Water supply and environmental management. Worldview Press

Okum DA (1988) The value of water supply and sanitation in development: an assessment. American Journal of Public Health 78 (11): 1463–1467

O'Meara M (1999) Exploring new vision for cities. In State of the World–1999

Rees WE (1992) Ecological footprints and appropriated carrying capacity: what urban economics leaves out. Environment and urbanization 4 (2): 121–130

Shiklomanov IA (ed) (1997) Assessment of water resources and water availability in the world. Comprehensive assessment of the freshwater resources of the world. Stockholm Environment Institute, Stockholm

Speets JA (1999) Poverty, health, water resources management in urban thirst and conflicts. Unpublished paper prepared for International Conference on Understanding the Earth System, 24–26 November 1999, Bonn, Germany

Sundblad K (1999) Regional Challenges. In: Falkenmark et. al. (eds) Water a reflection of land use: options for counteracting land and water mismanagement. Swedish Natural Science Research Council and UNESCO-IHP. Stockholm, pp 17–30

UNCHS (1996) Global urban indicators database. Nairobi

UNEP (1999) Global environmental outlook–2000 (GEO–2000). Geneva

UNESCAP (1990) State of the environment in Asia and the Pacific 1990. Bangkok

UNESCAP (1995) Urban water resources management: Water Resources series No. 72. Bangkok

United Nations (U.N.) Population Division (1995) World urbanization prospects: the 1994 revision. New York

United Nations (U.N.) Population Division (1997) World urbanization prospects: the 1996 revision. New York

Urban Age (1999a) The pipes are not leaking but the water 's gone. Urban Age 6 (3): 22

Urban Age (1999b) Access to clean water. Urban Age 7 (2): 29

Water Supply and Sanitation Collaborative Council (1999) Vision 21: A shared vision for hygiene, sanitation and water supply. Final Draft, 10 Dec

White RR (1992) The international transfer of urban technology: does the North have anything to offer for the global environmental crisis? Environment and Urbanization 4 (2): 109–120

WHO (1997) Health and environment in sustainable development. Geneva

WHO (1999) Water, sanitation and health at WHO-HQ. Website, //www.who.int/peh/

World Resources Institute (WRI) in collaboration with the United Nations Environmental Programme, the United Nations Development Programme, and the World Bank (1996) World resources report 1996–97: the urban environment. Oxford University Press New York

World Resources Institute (WRI) in collaboration with the United Nations Environmental Programme, the United Nations Development Programme, and the World Bank (1998) World resources report 1998–99: a guide to the global environment. Oxford University Press. New York

CHAPTER 16

Modeling Water Availability: Scaling Issues 16

Nick van de Giesen* · Luis J. Mata · Petra Döll · Arjen Hoekstra ·
Max Pfeffer · Jorge A. Ramirez

This report is the outcome of the workshop "Modeling water availability: Scaling issues" which took place on 25 November 1999 within the framework of the international conference "Understanding the Earth System: Compartments, Processes and Interactions", held in Bonn, Germany. Under chairmanship of Luis J. Mata, five presentations were given by scientists with different disciplinary backgrounds highlighting various aspects of water availability:

Jorge A. Ramirez, Hydrologist, Associate Professor of Civil Engineering, gave a three part presentation where issues of scale and dimensionality in geophysics were explored and highlighted. Two applications related to 1) the estimation of large scale water balances for the United States and 2) to the modeling of the continental scale dynamics of drought and soil moisture were presented which illustrate the complicating issues of scale and dimensionality in the modeling and prediction of the distribution of water availability.

Max Pfeffer, Professor of Sociology. This presentation illustrated the complexity of attempts to balance social responsibility and equity between communities of vastly different scales within the same watershed– 8 million New York City water consumers and the 250,000 people living in 60 rural communities supplying water to the City. The presentation focused on rural land use controls designed to be flexible in protecting the environment and providing for economic growth in rural communities. Specific land use controls often create diffuse costs in terms of reduced overall economic potential. New York City attempted to compensate for these costs by providing funds for infrastructure improvement, conservation, education, and economic development. This presentation argued that environmentally sound economic development can only be assured through exercise of an ethic or responsibility whereby rural residents weigh the downstream consequences of their local actions, and this outcome is encouraged through active participation in the watershed planning process.

Petra Döll, Hydrologist, "Water availability terminology, spatial averaging and criticality": Based on her experiences with global modeling of water availability and water use with the WaterGAP model (Döll et al. 1999), she first demonstrated the necessity of clearly defining the term "water availability" each time it is used as there are a few valid definitions (e.g. internal renewable resources vs. internal renewable resources plus inflow from upstream plus fossil groundwater).

---

* e-mail: nick@uni-bonn.de

The definition also must include the spatial averaging unit, as it is implicitly assumed that water can be easily distributed within the unit, e.g. the country, the river basin or the cell. Much work still needs to be done with respect to defining, at least at the continental or global scale, at what value water availability becomes critically low. Dr. Döll concluded that at these scales, the criticality of water availability would be best analyzed based on monthly values of water use (both consumptive and withdrawal use) and water availability (defined as internal renewable resources plus inflow from upstream) of river basins of about 50,000 km^2, for both long-term average climatic conditions and in typical dry years.

Nick van de Giesen, Hydrologist, gave a presentation under the title "Water availability in the Volta Basin". The aim of the presentation was to show to what extent the water availability as modeled at global scale by the WaterGAP model (see also Petra Döll's presentation) remains valid at the more detailed regional level and what type of adjustments are needed. Within the WaterGAP model, the Volta Basin is well-off with some parts having a water surplus and some parts being marginally vulnerable. Historically, however, the Volta Basin has had water crises in each of the past three decades. Reasons are vulnerability to a sequence of dry years (crises in eighties and seventies) and the larger than usual water need for hydropower (crisis in nineties). The size of the Volta Basin is 400,000 km^2 and 18 million people live there.

Arjen Hoekstra, civil engineer, presented a paper on 'managing water scarcity in the Zambezi basin'. The aim of this presentation was to show the significance of uncertainties in river basin planning. A new method for risk assessment in long-term planning is introduced. The basic idea behind this method is that uncertainties can be translated into alternative assumptions and that the risk of a certain management strategy, which has been developed under a given set of assumptions, can be analyzed by applying alternative assumptions. An assumption can refer to a particular parameter value, but also to the nature of a certain cause-effect mechanism as well. For the exploration of possible futures, the AQUA Zambezi Model is used as a supportive tool. Three 'utopias' and a number of 'dystopias' are considered. A utopia is based on a coherent set of assumptions with respect to world-view (how does the world function), management style (how do people respond) and context (exogenous developments). A dystopia evolves if some assumptions are taken differently. Using the risk assessment method described, the paper reflects on the water policy priorities proposed by the participants of a workshop in Harare in 1996. It is shown that putting the 'Harare priorities' into practice will work out effectively and without large trade-offs in only one out of the nine cases. It is concluded that minimizing risks would require a radical shift from supply towards demand policy.

The design of the workshop was to illustrate the path from detailed disciplinary research, through large scale integrative modeling, back down to regional scale management support:

First, two disciplinary studies concerning water availability in which issues of scale play a role; Ramirez representing geophysics and Pfeffer sociology.

Second, integration of social and physical information at a global scale with the Kassel WaterGAP model as state-of-the-art.

Third, two case studies at regional level, Zambezi and Volta basins, to sketch the difficulties which arise when social and physical science are integrated to support practical decision making.

In the ensuing discussion, it first seemed that the intended broad spectrum of perspectives had perhaps been too broad as it was hard to see the commonalties between the presentations. Although all scientists focused on scale, the approaches and terminology differed to such an extent that the actual issues at hand simply did not appear to be the same. Further analysis showed that this was partially due to fundamental differences between disciplines but also to a large part to confusion about "scale", "scaling", "aggregation", "integration", etc. For research which aims at integrating findings from different disciplines, it is necessary that there be some common understanding about the terminology involved with scale issues. One of the recommendations of the workshop therefore became the compilation of a "Practical guide to scale issues" which would explain what is meant by "scale" and related terms in different disciplines. At the plenary presentation, it was suggested to include such a guide in the workshop report. It was further remarked that work along these lines had been undertaken in the first IHDP working paper "Scaling Issues in the Social Sciences" (Gibson et al. 1998).

The remainder of this report should be seen as a step towards a guide which promotes cross-disciplinary understanding of scale issues. It is limited in the sense that it mainly discusses themes directly involving water availability. We start with definitions and descriptions of scale and dimensions. The discussion moves on to clarify scale issues in geophysics for non-specialists. Geophysics receives the bulk of the attention since Gibson et al. (1998) provides a fairly complete overview of scale issues in the social sciences. The most pertinent aspects of scale issues in the social sciences are briefly summarized. The report concludes with specific scale problems, and possible solutions, of integrative research. Throughout, material presented during the workshop will be drawn upon to provide examples.

Scale and dimensions

The term "scale" can take on many different meanings and for a consistent discussion we first need to state our definition. We will also say more about the related term "dimensions" but for now it suffices if length, time and mass are recognized as (fundamental) dimensions. To quantify an object, measurements are needed in all dimensions of interest. In order to make a measurement, a well-defined unit is needed such as a step, mile, day, karat, kilogram, etc. Measuring then becomes counting the number of times (including fractions) a unit fits the length, time interval or mass to be measured. The relatively clear difference between mathematics and physics is that the latter uses units which link theory to empirical facts. In some cases, scale is defined in a rather limited way as the device with which one measures such as a yardstick or a weight (Kroon 1990). Here, we will use a definition which better captures its common usage in global change research. The scale of a process or phenomenon is defined as its characteristic magnitude (length, duration, weight, etc.). This definition leaves room for imprecise scales such as global, regional, watershed, field, or point-scale but can also refer to more exact measurements such as the height of a wave or the duration of a rainstorm. In contrast to cartographic convention, large scale corresponds here to large characteristic magnitude.

Scale becomes an issue when data and/or process knowledge is available at a certain scale while predictions are wanted at a different scale in order to, say, manage water resources. In such cases, two basic phenomena may hinder moving from one scale to another; non-linearity and variability. If in Bonn, a weekend-long drizzly rain is measured by three rain gauges in the middle of an even plane, and all three gauges measure the same rain, one can safely assume that the rain which fell on the triangle formed by the gauges will indeed be very close to what the gauges measured. One can move from a small scale (gauge) to a larger scale (field). If, on the other hand, a tropical cloudburst fills one gauge but leaves one placed elsewhere in the watershed empty, the inherent variability of the process prevents us from knowing the total rainfall on the watershed. The variability problem may seem obvious but its solution usually is not.

Although every physical system can be quantified in terms of length, time, and mass, one normally uses derived dimensions such as electric charge, temperature, or even the social and economic dimensions. This is important when information is moved from small to large scale because usually one seeks not merely a collection of observations at the smaller scale but rather more comprehensive quantities which describe the behavior of the larger system as a whole. For example, urbanization in a given country can be described by enumerating the migratory moves of all individual families involved but for designing a water supply system one would rather know how fast the city grows and where.

For modeling system processes, it is convenient to describe the state of the system as a list of quantities and such a list is called the state vector. Each entry in the list (or each coordinate of the state vector) corresponds to a magnitude in a certain dimension. Complex systems will have many entries and are said to have a high dimensionality. If we return to the urbanization example, a complete state description could be a vector with a latitude and longitude coordinate for the position of each inhabitant. We may be fortunate, however, and find a useful lower dimensional vector, say a two-dimensional vector with only total area and population, which suffices for the problem at hand. Not only a larger spatial scale causes an increase in the number of dimensions but also integrating information from different disciplines which each bring in their own state vector. As we shall see, it is often increased dimensionality which hampers problem solving rather than larger scale as such when it comes to modeling water availability.

Scale issues in geophysical sciences

When geo-scientists (geologists, pedologists, hydrologists, meteorologists, etc.) say that a process scales, they actually mean that no characteristic magnitude is associated with the process but that instead there is independence of scale or scale invariance. The well-known fractals scale; one can zoom in or out but they always look the same. In geophysics, processes and variables usually have to be transformed or represented in a certain way before the scale-independence becomes clear. Considerable effort is put into finding out if and how processes scale because one can then subsequently limit expensive data collection to one scale and make predictions at other relevant scales. Ideally, one makes the equations and boundary conditions which govern a process dimensionless. The solution can then be multiplied with a scale factor of choice to obtain physical quantities again. Usually one is not so fortunate to have a tractable problem where all equations are known. In such cases, one may still find scaling properties empirically. To make this abstract story somewhat more tangible, the rainfall study by Ramirez will be presented in some detail.

Rainfall is measured by catching all rain falling in a funnel during a certain period of time. This period of time is usually one day for standard available rainfall data or, at best, one hour. For infiltration and erosion models, however, one needs to know how much falls in, say, five minutes or less. When rainfall is distributed evenly, there is no problem but especially during the more important cloudbursts, this is not the case and rainfall will vary wildly over short periods of time in a random fashion. The randomness, however, appears to have a scale-related structure:

$$E[R^n(\lambda A)] = \lambda^{\theta(n)} E[R^n(A)] \tag{1}$$

The following explanatory details are not essential for understanding the remainder but give an idea of what scaling exactly means in this case (for a good introduction, see also Gupta and Waymire 1998). R stands for rain, A and λA for the time (or space) intervals during (over) which rain is measured. λ is a so called scaling factor which in this case lies between zero and one. Therefore, the rain on the right-hand side is measured during an interval A, while the rain on the left-hand side is collected over a shorter interval, λA, between zero and A. Because we are working with random variables, we compare expectations (average values) denoted by $E[..]$. Actually, we compare the expectation of the rain to the power n, for both sampling intervals. The expectation of something to the power n is the so-called n-th moment. When n=1, we have the first moment better known as the mean, the second moment with n=2 is needed to calculate the variance $(=E[x^2]-E^2[x])$, and so forth. These moments describe the stochastic behavior of the variable at hand; rainfall, in our case. Empirically, we can calculate $E[R^n(\lambda A)]$ and $E[R^n(A)]$ and they happen to scale with a factor $\lambda^{\theta(n)}$, where $\theta(n)$ is a empirical function of n. If it always rains at the same rate, then $\theta(n)=n$; if we measure half as long ($\lambda=0.5$) we expect to measure half as much rain. For other stochastic variables following a simple scaling distribution, we may find $\theta(n)=n\theta$. For the cloudbursts of, in this case, New Mexico, we find the most complicated form of scale invariance called multi-scaling where $\theta(n)$ depends non-linearly on n.

Similar results have been found by sampling over different lengths or areas instead of over different time intervals. The importance is that when $\theta(n)$ is measured, one can deduce the statistical rainfall properties at all scales from the properties at one scale. Assuming $\theta(n)$ does not change, one can simulate fine scale (time and/or space) rainfall patterns on the basis of available coarse historical rainfall data.

When no relation such as in Eq 1 can be found which links observations at one scale to another scale the process is scale dependent. There is one characteristic magnitude at which the measurements and processes are valid. It may be possible to aggregate small scale observations to deduce behavior of larger systems. For example, one can apply mechanical statistics to find the gas law as some form of average behavior of many fast moving and colliding individual molecules and comparable approaches exist for modeling crowds and flocks. In modeling water availability, however, we often do not know how to calculate collective behavior when, again, non-linearity and variability play an important role. One could, of course, simply calculate the development and interactions of all small parts which make up the whole. For systems with a high dimensionality such as a natural watershed, this quickly becomes untraceable. Instead, the search is in first instance for low-dimensional descriptions of relevant large scale processes. Ramirez presented a good example concerning the actual and potential evapotranspiration at regional scale. Measurement and modeling of evapotranspiration (ET) at the point or field scale is rather complicated, involving the turbulent transport of water vapor and heat from the earth's surface into the atmosphere. It is common to distinguish potential (ETp) from actual evapotranspiration (ETa). ETp is determined by the atmosphere only and depends on wind-speed, air humidity, temperature, and solar radiation. The amount of water which would evaporate from a small open water surface defines in this case ETp. ETa is the actual amount of water which evaporates from the soil or is transpired by plants. At a point level, ETa depends on ETp but also on the amount of water available in the soil. The dependence of ETa on soil water is, however, hard to parameterize at small scales so one would have little hope to predict ETa for a region by summing the ETa calculated for every patch and field. Following Bouchet's hypothesis, however, a surprisingly simple complementarity relation between ETa and ETp was found for the conterminous United States east of the Rockies:

$$ETa + ETp = 2ETw$$

where ETw is a constant, equal to the evapotranspiration which takes place if ETa = ETp. What happens is that over large areas a high ETa reduces downwind temperature and moisture deficits and increased ETa may also result in increased rainfall within the region. If ETp is high, however, soil moisture is quickly depleted and ETa is reduced which increases temperature and moisture deficit, etc. This feedback mechanism, which would normally increase the complexity of the process, produced a symmetry which reduced the degrees of freedom to one!

Briefly summarizing, we see that geophysical processes may or may not have an associated scale or characteristic magnitude. In the case no characteristic magnitude exists, it may be possible to move data and knowledge from one scale to another, either by scaling the governing equations or by finding empirical rules. The issue becomes more difficult when a process does have a characteristic magnitude. In this case, qualitatively different behavior occurs at different scales as illustrated by the evapotranspiration study.

Scale issues in social sciences

As explained in the introduction, we will not go into much detail concerning scale issues in social sciences as these have been amply covered in Gibson et al. (1998). We would simply like to point out two, related, differences between geophysical and social sciences.

First, we would be hard pressed to provide an example of a social science issue concerned with water availability which scales in the geophysical sense of the word. In fact, all scaling issues brought forth by Gibson et al. (1998) would be characterized as non-scaling issues because all are scale dependent and have characteristic magnitudes (or levels, as Gibson et al. (1998) call them). In economy and ecology we do find scale invariance and it is surprising to see that the same multi-scaling which governs the rain in New Mexico also describes the price of IBM shares (Mandelbrot 1999). In general, however, such simplifying tools are not available when describing the human dimension of global change and resource management.

Second, when dealing with problems of scale in social issues, it is difficult to impossible to explain larger scale phenomena on the basis of the behavior of all constituents. Berger (1999) models the diffusion of new technologies in an irrigation scheme in Chile by simulating the behavior of 5,400 farmers but this seems, at present, a reasonable limit of complexity. We face here the same aggregation problem as in geophysical sciences that we can not trace the interactions of, say, 60,000 families in the New York City (NYC) watershed, let alone that of eight million New Yorkers. Instead, we again look for lower dimensional descriptions of the larger scale processes which are of interest. The main difference with geophysics, and this is methodologically probably quite crucial when findings from physical and social sciences have to be integrated, is that one can not forget lower and higher levels. Hierarchies figure large in Gibson et al. (1998), from ecology, through geography and economy to sociology. One level or scale in a hierarchy is to a large extent defined by the interactions with the levels above and below.

If we take the NYC watershed as an example we immediately see three important institutional levels, the farming communities in the watershed, the City of New York and the federal Environmental Protection Agency (EPA) (Pfeffer and Wagenet 1999). The City has for almost a century bullied the small rural communities in the Catskill mountains from where the City obtains an important part of its excellent drinking water without major treatment. In 1991, EPA issued the Surface Water Treatment Rule requiring communities to filter surface drinking water, a measure which would cost NYC 6 billion dollars in construction alone. The only way out is a strict watershed protection plan which should basically exclude the possibility of contamination. The farmers involved are mainly dairy farmers and cows are an important potential source of cryptosporidium which may cause disease if it enters the drinking water supply. Major changes in use of land and management of farms and rural communities are needed to protect the water sources in a way the EPA finds sufficient. At the same time, the City does not have much credit with the farmers and other rural residents. The present Memorandum of Agreement between the Coalition of Watershed Towns and NYC is comprehensive and contains important compensation measures but it remains unclear if all rural residents will take the kind of active responsibility for the environment which is needed for effective source protection. The analysis of the social and institutional dimensions of water availability for NYC can clearly not be understood without understanding processes at all three levels. This is different from physical sciences where large scale behavior of, say, gasses can be predicted on the basis of the simple gas laws without knowledge of the behavior of the constituting molecules. In the case of social phenomenon we must have an understanding of the motivations for environmental behaviors. These motivations relate to both individual-level preferences and community-level norms and expectations which limit individual actions.

Integrated science

Even if individual scientific disciplines would be able to solve scaling issues, integrating their findings will introduce, at least, two new problems. The first problem, described in some detail by Döll (Döll et al. 1999), is the complete mismatch of scale between social and physical data. Water availability is not just a question of how much water is there, but also of how much water a particular society needs. Knowledge of societal needs, e.g. of water demand, is derived from data which are typically generated and available at different administrative levels such as countries, provinces and towns. On the other hand, physical data, such as vegetation-cover and soil, are available as values of pixels with different variables usually having different pixel sizes and shapes. Riverflow and groundwater are most conveniently derived at watershed level while rivers actually tend to separate administrative units. Therefore, in integrated studies, there is the need to interpolate all data to a common spatial resolution. Due to the different spatial scales of the input data, the uncertainty of the model results is increased. Besides, it also becomes very difficult to assess the propagation of data uncertainties through the model.

The second issue, probably more crucial in the longer term, is that whatever hard fought success was booked within the disciplines to obtain a low dimensional description of large-scale phenomena, combining them will again cause an increase in the degrees of freedom, especially when dealing with watershed scales and larger. Water availability is a function of rainfall, groundwater, riverflow, demography, urbanization, agricultural and industrial development, etc. All these factors have to be considered over time and with all their interactions, in terms of quantity and quality. As Döll pointed out, this complexity is not only hard to manage but also makes effective communication with decision makers difficult. Water availability clearly has several dimensions but if a map for political decision makers has to be constructed one can usually only show one dimension and its spatial distribution. One can not show any subtleties such as, for example, that in one place, water is usually not scarce except on rare occasions when it becomes very scarce with particularly harsh social consequences whereas elsewhere, water is scarce regularly but that society is better adapted.

Scientific innovations in integrative research will come from reducing the dimensionality of the analysis while maintaining meaningful outcomes. Hoekstra presented what is probably one of the first of such attempts in the field of water resource management with the application of the AQUA model to the Zambezi basin (Hoekstra 1998). By applying cultural theory to water management in the Zambezi Basin, it became possible to combine physical data with social data and management styles for the construction of consistent scenarios. The idea is to start with a limited set of world views (in this case hierarchist, egalitarian, and individualist) and to work out so called utopias which are scenarios in which one world view is followed throughout. Observed facts and scientific projections are used as far as possible after which consistent choices are made with respect to uncertainties in the projections and possible interventions. For example, economic growth for the Zambezi basin was estimated by extrapolating present trends and choosing higher or lower values within the error margins. With a value 2.3 % from 1980 to 1991, the hierarchist scenario was based on 3 %, the egalitarian on 2 %, and the individualist on 4 % increase in gross national product per year. The end results are three scenarios which show likely future values and variance of water use and needs. As such, the approach is especially strong at pointing out the effects which different types of management have on water resources. Although, as Hoekstra himself pointed out, applying cultural theory has its limitations, it does provide a solution to the dimensionality issue which is clearly needed in any attempt at integrative research.

References

Berger T (1999) Innovationsprozesse und regionaler Agrarstrukturwandel. Ein Multi-Agentenmodell für Chile und den Mercosur. Schriften der GEWISOLA 35: 343–348

Döll P, Kaspar F, Alcamo J (1999) Computation of global water availability and water use at the scale of large drainage basins. Mathematische Geologie 4: 111–118

Gibson C, Ostrom E, Ahn TK (1998) Scaling issues in the social sciences. IHDP Working Paper No. 1, IHDP, Bonn

Gupta VK, Waymire EC (1998) Spatial variability and scale invariance in hydrologic regionalization. In: Sposito G (ed) Scale dependence and scale invariance in hydrology. Cambridge University Press, Cambridge pp 88–135

Hoekstra AY (1998) Perspectives on water – An integrated model-based exploration of the future. International Books, Utrecht

Kroon RP (1990) Dimensions. In: Besançon (ed) The encyclopedia of physics, 3rd edn Van Nostrand Reinhold, USA, pp 298–300

Mandelbrot BB (1999) A multifractal walk down Wall Street. Scientific American 280 (2): 70–73

Pfeffer MJ, Wagenet LP (1999) Planning for Environmental Responsibility and Equity: A Critical Appraisal of Rural/Urban Relations in the New York City Watershed. In: Lapping MB, Furuseth O (eds), Contested Countryside: The Rural Urban Fringe of North America. Brookfield-Ashgate, USA, pp 179–205

CHAPTER 17

Precipitation Variability and Food Security

Thomas E. Downing[1] · Fredrick K. Karanja[2] · Mohamed Saïd Karrouk[3] ·
Fred M. Zaal[4] · Mohamed A. Salih[5]

The break out group began with a roundtable discussion in response to the question: What are the key issues in understanding the linkage between precipitation variability and food security? The responses included:

Agriculture:
- Property rights
- Crop selection and rotations
- Agro-forestry
- Technology

Water:
- Intensive, efficient use in irrigation
- Participatory water basin management
- Long-term changes due to land use and climate
- Drought and desertification

Food security:
- Traditional means of coping in agrarian societies
- Global distribution of agriculture and globalisation
- Urbanisation

Uncertainty:
- Precipitation and nutrient flows
- Markets
- Spatial and temporal patterns

Research:
- Barriers to the integration of social and natural science

The group then had five presentations, on themes related to the above issues. A general discussion period followed the presentations. The first two presentations set the scene.

[1] e-mail: tom.downing@environmental-change.oxford.ac.uk
[2] e-mail: fkaranja@uonbi.ac.ke
[3] e-mail: karrouksaid@yahoo.com
[4] e-mail: a.f.m.zaal@frw.uva.nl
[5] e-mail: salih@iss.nl

What are the prospects for seasonal forecasts in Kenya?

Kenya has extremes of precipitation, ranging from deserts in the north and east to montane rainforests on the slopes of Mt Kenya. There are generally two rainy seasons – the long rains in March to May and the short rains in November to December. Their importance differs across the country, with the short rains being more reliable in western Kenya.

Agricultural activities are similarly dispersed – from nomadic pastoralism in the dry regions to commercial agriculture based on coffee, tea, grains, horticulture and other commercial crops. Most of the rural population depend on agriculture, either as smallholders, farm labourers or through secondary activities in the rural market towns.

In this developing country setting, the ability to predict the character of the forthcoming season could have enormous benefit. Potential beneficiaries of seasonal forecasting include:

- Subsistence farmers and pastoralists, who could anticipate food shortages through savings (either of food or cash)
- Commercial farmers, who could plan their inputs based on projected income.
- Labourers, who might find more or less work according to the expected harvests and may have to find alternative employment
- Agricultural planners, who monitor the national and regional food balance, schedule imports and exports and recommend interventions
- Relief and development agencies, who could gear up small-scale programmes or anticipate large scale assistance

Forecasts of seasonal climates in Kenya are still being developed. The basic mechanism is forcing of the atmosphere by the longer-term heat transfer from the oceans. Sea surface temperatures (SSTs) and sea level pressure (SLP) patterns are more predictable over the course of several months than the higher frequency changes in the atmosphere. Statistical associations indicate that the long rains in Kenya should be predictable based on SSTs and SLPs as part of the Indian Ocean/ western Pacific ENSO anomaly. Relationships for the short rains, which have more of an orientation from the Congo basin, are not as strong.

The most notable success in recent years was the prediction for above average rainfall in 1997, which resulted in widespread flooding. Government and local agencies were able to mobilise in advance of the flooding, providing some early relief efforts.

How are climate and food production related in Morocco?

Precipitation and temperature have changed during the last 100 years in many large-scale regions of the world, and not ably in Morocco. During this period, precipitation has varied considerably and variations have been relatively important. Variations in precipitation have accompanied strong fluctuations observed in the inter-tropical zone, especially the ENSO anomalies. These have affected temperature and precipitation in Morocco through the transfer of energy in the atmosphere. A negative anomaly, the El Niño, is manifested in Morocco by drought and a positive anomaly, the La Niña, by wet conditions. These extreme fluctuations in the Pacific have become more frequent in the last decade, and more severe. The recent droughts in Morocco have been particularly severe and catastrophic. Their brief interruption has been manifested by flooding. This increased instability of ocean-atmospheric events endangers water resources in Morocco, as well as agriculture and food security.

What can we learn about adaptation to climate change through coping with food insecurity during the past several decades of dessication in semi-arid areas of West Africa?

A Dutch project on the Impact of Climate Change on Drylands focuses on West Africa. Though the present trend is toward increasing rainfall, up till recently droughts were frequent and rainfall had decreased substantially. Global models of climate change indicate that the drylands of West Africa will be subject to lower and more variable precipitation, higher temperatures and higher evaporation. All contribute to higher risks of crop failure, and less food availability. Water, for humans and animals, and fuel wood may become scarcer too.

However, the impact of climate change on agricultural production and rural societies is speculative. This is not only because of uncertainties related to the magnitude and direction of climate change, but because agricultural production also depends on demographic, socio-economic and institutional factors, and on local decision makers' natural resource management strategies. Whatever the causes, less food, water and fuel locally available or produced means that the local population may be forced to adopt and develop alternative responses and that all kinds of institutions and organisations will have to adapt to a changing situation as well.

Adaptive options are available, and their relative importance for people may depend on the circumstances they find themselves in. Specific options may be combined into portfolios on the basis of past and present experiences (so-called pathways). However, some options may not be accessible for certain categories of people, thereby threatening their survival. The main objective of the on-going research presented in this paper is to determine the direction of climate change, its relative importance with respect to demograpic, socio-economic and institutional factors on agricultural production and the present and potential responses of the various actors on these changes.

The project so far has developed a spatial data base of agro-environmental conditions in West Africa. These have been related to different scenarios of climate change (see table 1). Some 40 % of the region is semi-arid or drier. In the Max Planck Institute (MPI) scenario, this is likely to increase, and especially so for the arid lands. However, the Princeton model (GFDL) shows little significant changes across the region. Clearly, climate change is uncertain, but potentially a significant risk.

Table 1. Changes in environmental conditions in West Africa with climate change

	1990	Climate Change Scenario, 2050	
		MPI	GFDL
Hyperarid	0 %	+/–	+/–
Arid	5 %	++	–
Semi-arid	35 %	+	+/–
Sub-humid	30 %	+/–	+/–
Humid	30 %	–	+/–

Key: ++   Strong increase, >50 %
       +   Moderate increase, >20 %
      +/–   o significant change
       –   derate decrease, <20 %
       —   Strong decrease, <50 %

Source: van der Born GJ, Schaeffer M, Leemans R (1999) Climate change on drylands with a focus on West Africa: Climate scenarios for the dryland regions in West Africa, from 1990 to 2050. Reports ICCD nr 2. Wageningen, CERES/DLO/WAU/RIVM

How will households cope with future food insecurity?

Global climate change is destined to produce global changes of a non-climatic nature. These changes are spatially multi-layered (traversing region, nation and locality), diverse (ranging from impaired environmental functions to food insecurity), cumulative (long-term consequences often outweigh the short-term ones) and integrative (not only climatic, but also social, economic and political). Six implications stem from this view of climate change and its global consequences:

1. Modeling of climate change and its impacts has tended to be global rather than local, and disciplinary and incremental, focusing on climate change, rather than integrating development into the climate change debate, and vice versa.
2. Locality-specific research on climate change and household food security has not kept pace with research on the global, regional and national levels.
3. The nexus between global development trends, climate change and changes in people's aspirations, attitudes and lifestyles is paradoxically problematic and cannot be understood by data extrapolation at the global or national levels isolated from local realities.
4. Household food security policies are more difficult for policy makers to deal with than national and global issues concerned with the quantity and quality of large economies than with disfranchised small-scale producers.
5. Given the current state of agricultural technology in most developing countries, changes in lifestyles and consumption patterns are indicative of potential pressures on the environment and possibly further environmental degradation. Climate change in such unfavorable circumstances will not make life easier.

6. Household vulnerability to food insecurity can be defined within a generalized notion of vulnerability, related to:
   - Household size and composition
   - Access to land
   - Enabling physical environment
   - Time availability
   - Knowledge or appropriate food processing and storage techniques
   - Income
   - Market access
   - **Expenditure pattern**
   - **Pricing policies**
   - **Power structure**
   - **Infrastructure and institutional capacity**

The range of coping **strategies includes (see table 2):**
   - Preventive strategies
   - Impact minimizing strategies
   - Creation and maintenance of labor power
   - Building up stores of food and salable assets
   - Diversification of the production strategy
   - Diversification of income sources
   - Development of social support networks

Strategies with a particular environmental component, and therefore subject to additional stresses through climate change, include use of wild foods, wetland cultivation, and horticulture.

Table 2. Matrix showing the dimensions and critical variables of food security

	Short-term 1–3 years	Medium-term 5–15 years	Long-term 25–30 years
Household	**Access to food** Nutrition and health	**Access to income** Improved well-being	**Improved incomes** **Social infrastructure**
National	**Safety nets** **Nutrition and health**	**Economic development** Rural profitability Agricultural research	**Food security for all** Poverty eliminated
Global	**Grain stocks** Food aid	**International research** Fair trading system Adequate global supplies	**International research** Fair trading system Adequate global supplies

Notes: Most important factors are in bold, and at top of list.
Source: after the World Bank.

How can we understand and measure vulnerability, particularly in the context of climate change?

A project with the UN Environment Programme has explored issues in vulnerability and adaptation to climate change impacts.[1] Along with mitigation of greenhouse gas (GHG) emissions, adaptation to climate change will be required, due to the long life span of GHGs in the atmosphere. A precautionary approach that plans effective adaptation strategies is now justified. This requires new tools of vulnerability assessment that provide guidance on coping capacity and adaptation. Both vulnerability and adaptation are priorities in the UN Framework Convention on Climate Change (UNFCCC). Three key questions are explored below.

Why do we need information on vulnerability and adaptation?

Due to the long life span of greenhouse gases in the atmosphere, and regardless of efforts to reduce their emissions, climate change will occur at least through the middle of the next Century. Adaptation is a necessity, and a precautionary approach is justified.

Planning effective adaptation must begin with a clear understanding of vulnerable populations and regions based on an assessment of the capacity of these groups to cope with climate variability and change. Some communities recover from hurricanes, fires, floods, and other extreme events more rapidly and completely than others because coping and adaptation strategies are not equally available to all affected populations. Certain strategies require institutional infrastructures (e.g. agriculture extension services, and insurance markets); others require public expenditures (e.g. flood control, disaster relief, and subsidized disaster insurance); others (e.g. informal income support) benefit from the presence of tightly knit communities.

Vulnerability and adaptation are central to international policy on climate change, in both the UNFCCC and the Kyoto Protocol. Phase II adaptation actions are being considered. Funds from schemes such as the CDM will need to be carefully targeted.

Vulnerability is a function of sensitivity to present climatic variability, the risk of adverse future climate change and capacity to adapt.

Many of the impacts of climate change are uncertain; equally, vulnerability is dynamic, continually changing as a result of various factors. Therefore, adaptation strategies should promote responsive institutions that maintain a rich repertoire of policy options. Social resilience in the face of climate change should complement the aims of sustainable development.

Composite vulnerability indices would be useful tools to help identify vulnerable situations and plan and monitor effective adaptation measures.

Achieving agreed composite indices requires an international effort, recognising the complex issues and challenges (see fig. 1).

---

[1] This summary draws upon the project summary presented to the UNFCCC Conference of Parties in Bonn, November 1999.

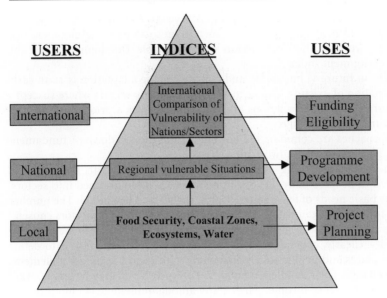

Fig. 1. Users and uses of vulnerability indices

What are the issues and challenges?

An example may help frame the issues and challenges. In the case of food security, some regions are more sensitive to climate change than others – for example, semi-arid areas or coastal regions subject to flooding and saline intrusion. The capacity to cope also varies – some socio-economic groups may not have access to new agricultural technology, conversely, some regions will be able to develop additional irrigation. The impact of climate change on food security will be conditioned by the both the sensitivity of food systems to climatic disturbances (such as drought) and the coping capacity of peoples in sensitive regions.

A quantitative assessment of vulnerability needs to define indicators of sensitivity and adaptability, in this example for food security and climate change. It is this search for quantitative indicators that distinguishes this effort from current climate impacts studies.

A clear definition of what the indices will be used for is essential. The above figure suggests a variety of users and uses for quantitative indices of vulnerability. At the global level, composite index could determine eligibility for funds to reduce vulnerability and encourage adaptation to climate change. At the local level, sectoral indices could be instrumental in designing and targeting projects. At the national (or regional) level, vulnerability indices would aid in planning adaptation strategies. The scale for determining vulnerability indices being a function of the intended purpose of the indicator and the availability of data. Monitoring vulnerability at all levels would help create awareness about future climate change impacts.

Formal vulnerability assessments are well developed in some sectors. Insight can be gleaned from existing indicators, such as on sustainable development (for example the work of the Commission on Sustainable Development) and the Human Development Index.

Research on future vulnerability and climate change adaptation is at an early stage. Evaluation of existing studies may help to identify sectors where susceptibility to current climatic variations and hazards indicates a priority for adaptation to future climate change.

Two approaches are common: the bottom up and the top down. A fundamental challenge is to link these two scales.

Bottom-up understanding of local vulnerability begins with the pressures experienced in people's daily life. Such pressures are often grouped into sectors, such as the basic needs of food, water, health, shelter and livelihood. The number of sectors to be included must be determined and appropriate indicators chosen.

The top down approach seeks to compare countries according to their relative vulnerability. Clearly, countries per se are not vulnerable – the task is to define vulnerable places and populations, particularly with regard to different elements of climate change. For example, an index of GDP per capita may be less relevant than the distribution of income, particularly among vulnerable communities.

The bottom-up, sectoral approach needs to be scaled up to represent vulnerable situations. For example, the sensitivity of marine ecosystems and coastal settlements needs to capture the vulnerability of fragile fisheries and coastal economies. Defining regional vulnerable "syndromes" and combining indicators at this intermediate level is a major challenge.

The geographic scale must be clarified. For small countries, a regional scale may be tractable. For example, patterns of vulnerability are likely to be similar throughout the Caribbean. For large countries, analysis of sub-national regions may be appropriate.

Integrating across sectors and comparing situations is a major challenge. For example, one approach is to count the number of indicators that exceed a certain threshold, flagging that indicator as contributing to vulnerability. While this is easy to calculate, interpreting the significance of each flag and their cumulative importance is less readily accomplished.

Another challenge is distinguishing between present vulnerability and the additional vulnerability related to climate change. This may be a major concern if international funds for climate change adaptation are to be targeted at the additional risk rather than reducing present vulnerability, for example by promoting sustainable development.

Above all, the methodology and judgements must be transparent. Independent audits of indices by impartial experts may be desirable.

The quality and quantity of data required is a concern. There are standard sources at the national level, but these may need to be further evaluated. Much data is currently available, even at the local level, and programs for data standards and reporting are operational. Expanding the national communications could be useful to report data for more formal vulnerability assessments.

Comparing national vulnerability is a contentious issue. It may be desirable to seek agreement on criteria for vulnerability indices, in much the same way that the IPCC provides advice on climate change scenarios but does not recommend individual model simulations.

The difference between situations can be enormous based on local adaptation capacity. The quest for quantitative numbers should not hide the disparity between, say, the capacity to adapt to sea level rise in The Netherlands, mainly a question of economic protection, and developing countries where sea level rise threatens vulnerable livelihoods.

Ultimately vulnerability assessment seeks to provide insight into effective adaptation strategies. The analyst and the user must set thresholds, group indicators and interpret indices. A composite indicator of vulnerability is a device for analysis rather than something "out there" that can be objectively measured.

What is the way forward?

Capacity building in vulnerability assessment, contributing to an international process to develop quantitative indices of vulnerability, is required. The process is as important as specific gaps to be filled. UNEP is seeking input regarding this process and specific proposals.

The process must ensure that the resulting methodologies meet the needs of the UNFCCC, address the commitments of subsequent protocols and help advance work on Stage II Adaptation in the GEF.

Participation in the process must be international and global. Parties, their scientific advisors, international agencies, nongovernmental organizations and experts need to recognize the collective process and secure its ownership. In its essence, vulnerability bridges between the poorest and wealthiest countries; both need to understand the issues and contribute to the process.

A network of international research community needs to be developed. Vulnerability spans many disciplines; vulnerable situations occur in all countries. At present, research on vulnerability and especially formal indicators, is fragmented.

Analytical frameworks and options for implementing composite vulnerability at different levels need to be evaluated. Methods should be included in the IPCC reviews. Developing further guidance for vulnerability assessment in national communications may be warranted.

The process is going to take time. The scientific and technical tasks are significant. The need to consider societal values and political processes requires time to develop and review proposals.

A pilot phase of testing and implementation is essential before recommending methodologies. The pilot phase should enlist users in the process and demonstrate specific applications.

Specific research gaps include: linking local, sectoral assessments and national composite indices; understanding the cumulative effects of additional stresses caused by climate change; providing profiles of vulnerability that are relevant to diverse users and uses; and validating assessments.

CHAPTER 18

Interdisciplinary Perspectives on Freshwater: Availability, Quality, and Allocation

Andreas M. Ernst[1] · Wolfram Mauser[2] · Stephan Kempe[3]

Freshwater is a fundamental resource and a basic requirement for life on Earth: It is the unifying agent of natural ecosystems, water circulation (e.g. Baumgartner and Reichel 1975) plays a central role in the global cycles of elements (e.g. Degens et al. 1991) and is directly linked to climate. Water is a prerequisite for biomass production which is often limited by local water availability. To man, water not only secures food supply but its quality is also the basis for human health (e.g. Meybeck et al. 1989). Its availability also enables economic development, and thus constitutes a possible cause of political conflicts. Even within societies, water is subject to competing usages (e.g. for irrigation, transportation, energy production, recreation, drinking water and others).

On the other hand, water on the land surface is increasingly and intensively influenced by human activities and decisions. Since water is a crucial substance in the causality chains producing biodiversity disturbances, it not only plays a pivotal role in the natural physical, chemical, and biological processes, but also in the anthropogenic changes to the Earth System. Water related risks concerning its availability, quality and allocation affect a major part of the global population and thus influence human health as well as safety.

There is a growing consensus in the international discussion on global change and the concept of sustainability that securing the availability and quality and adequate allocation of water is among the most important issues of the future. In Germany the Scientific Advisory Board to the Government on Global Environmental Change (WBGU) dedicated its 1997 annual report to this issue (WBGU 1998). Still, efficient water management (in terms of both quantity and quality) barely exists in most parts of the globe and needs adequate, advanced and cost-efficient technology. Much of the necessary transitions towards a sustainable management of resources are thus related to water availability, water quality, water allocation and their determining factors.

An international workshop was held from November 26. – 28. 1999, at the Stresemann-Institute, Bonn, Germany, uniting experts from Europe and Africa in the fields of hydrology, water resources management, social sciences, ecology,

[1] e-mail: ernst@psychologie.uni-freiburg.de
[2] e-mail: w.mauser@iggf.geo.uni-muenchen.de
[3] e-mail: kempe@bio.tu-darmstadt.de

and meteorology. The topic of the workshop was research on sustainable water management within the global change context with special emphasis on "Freshwater – Challenges for Research and Policies: European Response to a Global Problem". The German National Committee on Global Change Research (NKGCF) and the German Federal Minister for Education and Research (BMBF) supported the meeting.

The theme of the workshop closely relates to an integrated national global change research project on "Water: *Availability, Quality and Allocation (AQuA)*". Its instantiation was recommended by the German National Committee for Global Change Research and realized by the German Federal Ministry for Education and Research (BMBF) under the title "Globaler Wandel des Wasserkreislaufs" (*Glo*bal change of the *wa*ter cycle, *GLOWA*). The program aims at (a) the development of techniques for integrated modelling and decision, (b) the development of technologies for sustainable water resources management in different ecosystems, cultures, and societies including flood control, drought prevention, irrigation, water quality management, (c) the development of technologies to establish a (ground-based and remote sensing) monitoring system for water availability, (d) the establishment of human capacities in decentralised competence centres in the field of integrated water research, and (e) the strengthening of co-operation between the social sciences and natural sciences.

The aim of the workshop was to initiate international co-operation by bringing together national and international expertise in all fields of water related research with special emphasis on Europe and Africa. Through plenary meetings and working groups, the workshop pursued the purpose to (1) identify a set of high priority research activities which should be initiated to improve methodologies of integrated water research in watersheds of different climatic and cultural conditions, and (2) discuss new approaches to integrative water research and modelling with the aim of a sustainable future management of water resources and thus (3) open up perspectives on the integration of scientific disciplines on the environmental and socio-economic dimensions, including issues of the developing countries as well.

This paper resumes the central results of the workshop[1]. It presents the most relevant research issues identified. They concern water allocation conflicts, water quality issues, global and anthropogenic change, and possible environmental protection measures. It discusses methodological issues (like scaling and predictability problems), and gives a perspective on interdisciplinary work.

Most relevant issues

The first aim set for the workshop was to identify the most important freshwater research issues and problems. A special emphasis was given to issues that were considered only to be solvable in a multi-disciplinary way. In the following sec-

---

[1] We thank all the workshop participants for their contributions, especially the rapporteurs and chairs of the working groups.

tions, these issues will be presented grouped around the topics of water allocation conflicts, quality issues, global environmental change and anthropogenic factors, and environmental protection strategies.

Water allocation conflicts

Water allocation, i.e. the distribution of freshwater resources among users, is implemented through informal or institutional decision making and regulations. They aim at managing existing freshwater demands given a certain freshwater availability. Central to this management is the goal to equilibrate the demands of the competing agricultural, private, industrial and also natural sectors. Conflicts may arise because of multiple water usage for drinking and sanitation purposes, irrigation, industrial processes such as the manufacturing of goods, mining, cooling, hydropower generation, shipping, recreation, or sewage disposal. An additional complication may be caused by competition among nations. Water allocation thus possesses spatial as well as temporal aspects (cf. Ernst 1999a).

When considering distribution of freshwater resources, managing water *scarcity* is the first thing to come to mind. Indeed, water scarcity caused by climatic variability presents the main problem in semi-arid watersheds. It is of utmost importance and has serious consequences for everyday life including food and water supply of the population and their cattle. Water recycling, as discussed in the section on environmental protection measures, constitutes a possible technical option to mitigate the water scarcity allocation problem.

In temperate lowland watersheds like those of the Elbe, Oder, Rhine, or parts of the Danube in Central Europe, a decreasing water availability does not lead to freshwater scarcity per se. But still, upstream-downstream conflicts exist and mostly centre on quality concerns downstream.

In mountainous watersheds, water allocation problems relate to a local or seasonal water *surplus* to be distributed in order to prevent floods in the lowlands. This problem is most often dealt with in an exclusively technical way by building dams to lower discharge peaks. This approach alone, however, can not solve the flood problem, as newer research results show. An example from northern Sweden may serve to illustrate the point (Bergström and Lindström 1999). The flood protection given by the dams leads to a considerable risk reduction during the high or medium return rate floods (i.e. events in the range of 10 to 30 years). This in turn has fostered a land use change: Regions in the downstream flooding areas have been populated and developed, but the dams cannot handle more extreme events with longer return rates. These events consequently cause very high damages in the now developed areas. It seems that a mostly successful flood protection management makes people feel too secure and reduces their awareness of the less frequent, but still remaining high risk events. There exists evidence that the integral of damage remains the same or even has increased compared to times before the regulation of the rivers. Such argumentation is also true for large flood plain rivers, which suffer climate induced century floods, like in the case of the 1982/83 El Niño flood of the Rio Paraná (Depetris and Kempe 1990).

Further problems of reservoir operations are related to conflicts between the need of continuous energy production upstream and water usage by fishing or tourism downstream, which expect a more natural flow regime. A balance thus has to be negotiated between the different functions of mountainous regions, also for their surrounding lowlands.

Water quality issues

As was already pointed out in the introduction, water plays a central role in connecting physical, chemical and all biological earth processes. As such, its quality is crucial to a wide variety of life and also especially vulnerable to disturbances. Unwanted influences can act on a whole set of quality factors: Temperature has effects on gas concentrations and flora and fauna, major ions influence the salinity of the water, and lead to concentration problems for drinking water, minor ions have toxic effects. A surplus of nutrients like P and N and a decrease of Si lead to eutrophication, anaerobia, unwanted blooms, and again cause problems for drinking water. Gases like H_2S, O_2, CO_2 may be enriched or depleted and (e.g. Kempe 1982) lead to unwanted smell and fish kills. Heavy metals and chlorinated hydrocarbons have toxic effects. A surplus of organic carbon is the origin of oxygen deficiency and excessive sediment formation, bacteria cause fish kills and represent a risk to human health (causing e.g. tropical diseases).

These constituents and properties of freshwater are subject to long-term trends, seasonal changes, addition and withdrawal of water, impoundments, tributary mixing, and general downstream evolution. It becomes obvious how sensitive freshwater resources may react to natural and anthropogenic influences. The implementation of water quality monitoring systems in large parts of the globe is still needed to serve as the basis for sound water management decisions to allow for development and to prevent a further degradation of the resources. However, as will be shown below, the monitoring has to be complemented by a discussion on the habits of water use, and reuse, leading to quality target values that are widely accepted within a society.

Global and anthropogenic change

The interaction of human activity towards water with global environmental change is expected to have very distinct effects on different regions due to different land and water uses, human lifestyles and therefore is expected to entail a variety of different vulnerabilities.

Though the wetlands are also vulnerable to global climate change with respect to nature conservation and biodiversity issues of flora and fauna, the most salient human influence on fresh water and its availability in the lowlands is the agricultural land use. It can be concluded that, for the lowlands, regional impacts on water seem to be of higher concern than global climate change. Global change, however, will cause shifts in water quality, will alter temperature impact on river biota, and will change dilution capacity and other factors.

In the temperate mountains, a somewhat higher impact of even small global environmental changes is expected. Mountains possess a higher vulnerability because of large inputs of water and energy, and larger stresses may lead to more important changes in runoff. Direct anthropogenic changes in the utilisation of water resources in the mountains result from mining, timbering and hydropower production (which is, e.g., providing 70 % of the energy production in Austria), and tourism (e.g. in the Alps).

Problems of global environmental change, however, turn out to be especially pressing in semi-arid regions. Population growth has reached an upper limit, which has resulted in extremely reduced ecosystems with only few species. Land degradation, erosion caused by wind, salination and pollution of the resources are only a few of the consequences. Diseases are caused by water scarcity and can have serious impacts on the population of semi-arid regions. In general, the nutrient cycle is closely linked with the water cycle and thus with global change. Nomadic lifestyles and migration are influenced by environmental stresses. Both environmental variables and the human reactions to them possess a high variability and are long-term processes. This adds to the difficulties in predicting and modelling them, and in finding acceptable ways of mitigation.

We can conclude from the reported observations, that the regions mentioned above each are characterised by differential vulnerabilities to global environmental changes. These vulnerabilities closely relate to human activities and lifestyles and the expected human adaptation to the environmental changes.

With respect to sustainability of fresh water demands, the workshop participants discussed possible strategies aiming at the protection of environmental systems. Traditionally, environmental protection is effected through emission control measures, i.e. regulations limiting the release of pollutants from emittants into the environment. More recent initiatives of legislation, e.g. the EU water framework directive, also stress immission control as a second part of a combined water quality strategy (see fig. 1). The major objective of immission control strategy is to reduce the environmental impact to (or below) critical levels (i.e. quality target values). These target values have to be derived on the basis of ecological and eco-toxicological investigations, but they also touch on the social perception and acceptance of risks.

The dependence of the target values on the water usage also allows a multiple use of water. Here, the water quality standards to be met may be reduced stepwise

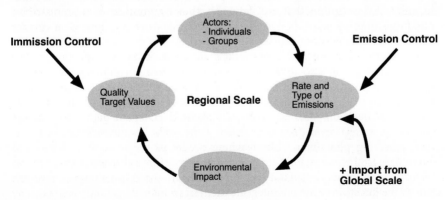

Fig. 1. Two strategies of protection of environmental systems

from one usage to the next. The resulting "recycling cascade" may serve as an intermediate step towards a closed recycling system in which water is totally re-used and the amount of waste water reduced to a minimum.

It has to be noted, e.g. in semi-arid regions, that poverty simply might prevent an easy implementation of possible (e.g. technically and socially costly) solutions because of limited resources. This relates to the more general problem of fitting cultural needs and resources to technical and political solutions, or, phrased differently, it relates to the acceptance problem.

Methodological issues

Apart from discussing the main water specific problems relating to the different regions of the world, the workshop also tackled important methodological issues of water research. Among these, the issue of connecting views on different scales, the handling of forecast problems, and some experiences and recommendations on interdisciplinary working are reported here.

Scaling

Comparing the temporal as well as the spatial scales of description and prediction used in the different disciplines working in the field of water research, their large variance immediately becomes obvious.

Meteorology, e.g., mainly uses macro-scale and nested regional models with a time scale of 25 years or more with respect to global change scenarios. A hydrologist would work on point-, micro-, or meso-scales with relating time step widths, depending on the task at hand. A region that lends itself to consideration from the point of view of hydrology would be a river basin. Ecology and agricultural sciences possess a similar flexibility, depending on their research task. Their spatial scales range from 1 ha to 1 km^2 (ecology) or even to several 100 km^2 (land use considerations), their temporal scales ranging from hourly (ecology) to daily or yearly time steps.

Some social sciences, like economics, political science, and sociology, use spatial units of description that are defined rather by political or administrative boundaries than by natural ones. These boundaries neither follow water divides (rather water divides often define political borders) nor do they coincide with the rather practically motivated geographical information system quadrants. The effects of changes on a societal level (as a reaction to environmental or political changes) can often only be observed on larger time scales (i.e. > 20 years). Psychology, on the other hand, mostly looks at actors (such as persons, households, institutional decision makers, or other stakeholders, again depending on the research question) or at groups of them. Time scale here sometimes is short-term, reflecting phenomena like learning processes or decision making, and sometimes also including long-term learning and attitude changes (Ernst 1999b).

The wide spectrum of spatial and temporal scales used in freshwater research leads to the question of how to integrate them in an interdisciplinary context. The different disciplines can be considered as working on a number of systems that are nested within each other. Freshwater quality factors, for example, are nested in global driving forces (like shifts in climate patterns, i.e. precipitation and temperature) investigated by meteorology, and global imports and exports (specifically for water and nutrients) linked to ecology, social sciences, and others. Likewise, individual decision processes are nested in economical and political constraints, defining them at the same time to a certain degree. Nevertheless, the question remains of how exactly to couple these nested models of different grid sizes on spatial as well as temporal scales. It can be argued that the different scales may be bridged through a more detailed analysis in pilot sites and subsequently the development of translation mechanisms to be applied to the regional scale at a later stage. That means one has to look for simpler algorithms that describe the detailed picture found in the pilot sites with sufficient precision, a task that may well be one of the key challenges of the fresh water research for the next few years. First steps in this direction have already been made.

Modelling: Predictability and scenarios

Integrated modelling of the economic, sociological, political and availability aspects of water was considered by the workshop participants to be an important tool. It helps reaching reasonable, justifiable and feasible solutions to fresh water quality and allocation problems. Another useful methodological development would then be the construction of decision making tools based on the integrated models. They could be used by those responsible for water quality management and water distribution.

However, obtaining scientifically sound predictions from such models may turn out to be a far too ambitious goal to reach. In various areas one can show that such predictions neither seem possible now nor in the nearer future, concerning, e.g., the retention potential of lowland watersheds or their uncertain flow directions, the rainfall in temperate mountainous watersheds, the adaptation of natural systems to global change, or societal reactions to global change and the societal impacts on different levels (with regard to such diverse factors as water quality, habitat loss, employment rate etc.). Predictions become especially difficult to make when not only direct effects but also indirect feed-backs have to be considered.

One way to go is to develop and to carefully analyse and interpret scenarios based on running models and on a set of constraining variables. These scenarios could give good impressions of different simulated courses of action under a variety of natural and societal conditions. However, it is necessary to consider the high degrees of freedom immanent to all scenarios, especially those including long-term perspectives. On the other hand, safe action corridors and areas of resilience may well be identified through scenarios, which may give clues to what course of action is the most promising. An important step will be the scientific and political evaluation of the scenario outcomes in relation to coping with global environmental changes. It may be best implemented by an agreed standardised procedure, comparable to the "benchmarking" known from computer hardware development. For example, consideration would not only be given to economical and social indicators of the scenarios, but also to how much degree of freedom remains for society to react to the challenges in the future.

Interdisciplinarity

Water use and allocation call for interdisciplinary research approaches. This is not only true for the collaboration within the natural sciences, but to a substantial degree also for the collaboration of the natural sciences with the social sciences. Two examples are given to illustrate the point.

The Nile is an example of a multiple-used and therefore highly impacted system. The completion of the Asswan Dam in 1964 has protected Egypt from the annual Nile floods, has provided enough water throughout a drought spell in the eighties and has made triple cropping possible in Egypt. On the other hand, salinization is the consequence: At the dam site the Nile throughput is approx. 55 km^3/year, carrying 13 mg Cl/l on average. At Edfina, only 4.5 km^3/year are monitored, carrying 33 mg Cl/l. That is not enough to maintain the salt balance. One can calculate that roughly 570,000 t of chloride are lost in the system amounting to a salinization of 0.4 t NaCl/ha/year. This is enough to bring a 1m thick layer of groundwater equalling the agricultural area of Egypt to sea water salinity in 30 years (Kempe 1993). Once these hydrological facts are known, the research questions start to touch on societal practices, e.g. how can the current practice of irrigation be changed to increase the salt-output to the Mediterranean Sea to a value allowing long-term usage of the Nile Valley and hence sustainable development? Research to answer such questions would involve hydrochemistry, hydrology, agricultural sciences, and social sciences like sociology, cultural anthropology, psychology and economics. In this case, it seems essential to understand the system with its feed-backs on the different levels involved, since solving one of the water related problems may cause others.

As another example, the nitrogen cycle nicely illustrates the need for an interdisciplinary perspective. This cycle involves many aspects of the water cycle and is of local, regional and global importance. It is governed by a multitude of natural, environmental and anthropogenic factors, for example: global change (e.g. the impact of global climate change), population growth in connection with an increase of standards of living, industrial development including traffic, and agriculture and cattle raising in connection with land use change and changes in fertilisation practices. In general, it can be noted that the target values of environmental interventions strongly depend on ecosystem functions and the type of water use. In order to meet these targets socio-economic instruments have also to be adopted which aim at reducing the pollutant release of the different emittants (individual actors and groups). Furthermore, cost-effectiveness and acceptance of the measures must be taken into consideration when trying to change the habits of the emittants. Within this cycle between emission and environmental impact, the natural sciences traditionally have focussed on developing tools (numerical models) to simulate and predict the fate of the environmental factors. Their models can be used to show where the cost-effectiveness of the socio-economic measures is maximal to meet the targets of living quality. The cycle just described demonstrates again the close interplay between social sciences and natural sciences. More generally speaking, *the challenge is to link the analysis of the natural world with institutional analysis, the analysis of the policy environment and the analysis of individual decision making.*

Table 1. Types of knowledge and the coupling of sciences in interdisciplinary water research

↓	Process Knowledge	Natural Sciences
	System Knowledge	Natural + Social Sciences
	Target Knowledge	Social + Natural Sciences
	Transformation Knowledge	Social Sciences + Technology

It may be helpful to characterise the different types of knowledge involved in such a project by using a hierarchy with increasing social science background (see table 1): While the natural sciences lay the ground describing the natural processes, an understanding of both natural and social factors is needed to adequately model environmental systems with their anthropogenic driving forces (like the Nile in the example above). When it comes to defining environmental target values in relation to societal needs, also both sciences have to contribute their knowledge. Showing ways of transforming the existing state of affairs into the desired one finally is a task for the social sciences, supplemented with technological solutions, where possible. Any mitigation technology and intervention has to be adapted to the local circumstances. Though many problems seem to be similar on a technical level, the solutions may be different due to differences in the socio-economic, cultural and institutional settings in which they are embedded.

We illustrated what can be expected from a cross-disciplinary collaboration in water research. However, the process of bringing together the disciplines in a truly common understanding of the domain and its main problems should not be ignored. Observation of the following principles should foster interdisciplinary research:

(1) Do not *a priori* define the interaction among the disciplines in terms of deliverables.
(2) Orient to and concentrate on small definable problems.
(3) Define those problems together.
(4) Derive deliverables from the problems together.
(5) Mutually trust collegial expertise.
(6) Organise feed-backs (i.e.: does the big interdisciplinary picture cover my discipline's contribution adequately?)

In this context, funding agencies are encouraged to allow for new approaches of project organisation. Though accountability must be ascertained, the interdisciplinary work groups must be given a chance to organise themselves around both a common understanding of their problem and possible solution strategies.

During the workshop on freshwater research, participants agreed that integrative approaches are necessary in this field of research. Even though significant differences are noticeable in the ways of tackling problems and in the underlying research philosophies between the natural and the social sciences, they both become aware of their complementary potentials and roles in an integrated research process. It is important to bear in mind that the natural sciences define the framework within which human action takes place but that they cannot define human action itself. Finally, it seems to be an especially constructive starting point to be aware of the uncertainties in our knowledge in any discipline and the limits of our forecasts. This will lead to relaxed and fruitful relations between the disciplines, also in water research, and bring their collaboration to bear.

References

Baumgartner A, Reichel E (1975) The World Water Balance. Oldenbourg, München

Bergström S, Lindström G (1999) Floods in regulated rivers and physical planning in Sweden. Talk given at the ICOLD Workshop on benefits of and concerns about dams – case studies, September 24, 1999, Antalya, Turkey

Degens ET, Kempe S, Richey JE (eds) (1991) Biogeochemistry of Major World Rivers. SCOPE Report 42, Wiley & Sons, Chichester

Depetris PJ, Kempe S (1990) The Impact of the El Niño 1982 event on the Paraná River, its discharge and carbon transport. Palaeogeography, Paleoclimatology, Paleoecology (Global and Planetary Change Section) 89: 239–244

Ernst AM (1999a) The psychology of environmental action. In: Barrage A, Edelmann X (eds) Recovery, Recycling, Re-integration. Collected papers of the R'99 World Congress Geneva Switzerland (Volume I). EMPA, St. Gallen, pp 3–7.

Ernst AM (1999b) Resource dilemmas, computer simulated actors and climate change – A methodology to investigate human behavior in a complex domain. In: Stuhler EA, de Tombe D (eds) Complex problem solving – Cognitive psychological issues and environmental policy applications. Hampp Press, München, pp 95–105

Kempe S (1982) Long-term records of CO_2 pressure fluctuations in fresh waters. Habilitationsschrift. In: Degens ET (ed) Transport of Carbon and Minerals in Major World Rivers, Pt. 1, Mitt. Geol.-Paläont. Inst. Univ. Hamburg, SCOPE/UNEP Sonderband 52: 91–332

Kempe S (1993) Damming the Nile. In: Kempe S, Eisma D, Degens ET (eds) Transport of Carbon and Minerals in Rivers, Lakes, Estuaries and Coastal Seas, Pt. 6, Mitt. Geol.-Paläont. Inst. Univ. Hamburg, SCOPE/UNEP Sonderband 74: 81–114

Meybeck M, Chapman D, Helmer R, (eds) (1989) Global Freshwater Quality, A First Assessment. Global Environmental Monitoring System. WHO, UNEP. Blackwell, Oxford

WBGU (1998) World in transition: Ways towards sustainable management of freshwater resources. 1997 annual report. Springer, Berlin

CHAPTER 19

Water Deficiency and Desertification

PAUL L.G. VLEK* · DANIEL HILLEL · J.C. KATYAL · WOLFGANG SEILER

In a historical perspective, societies seem to have forever struggled to manage their natural resources. Failure to do so effectively and sustainably, ultimately led to the demise of cultures and the displacement of populations (Hillel 1992), an occurrence known today as "environmental refugees". At stake were largely the resources water and land, with close coupling between them. Water is becoming increasingly scarce (Falkenmark 1997). Not only are the demands placed on water by urban and industrial centers competing with its traditional users, the farmers, but these new type of users are often polluting the water to such an extent that it is not readily reusable. The situation is far from stabilized, with urbanization continuing at unprecedented levels and the effects of fossil energy dependence adding to greenhouse gas accumulation in the atmosphere and to the vagaries of the weather. The working group on "Water Deficiency and Desertification", with about 25 participants, discussed these issues based on three short introductions by the authors.

Climate change

The Institute for Atmospheric Environmental Research (IFU) in Garmisch Partenkirchen, Germany, has identified Climate Change as the primary environmental issue (consistent with the GEO 2000 survey by UNEP) affecting the future management of land and the danger of its degradation. Some of the problems associated with climate change are the deterioration of air quality, ozone depletion, and desertification. IFU has demonstrated the need for forecasting climate change on a regional scale in order to provide a basis for scenario testing and as a means of early warning. The soil-vegetation-atmosphere feedbacks are too intricate to be captured by the existing global climate models (GCMs). Regional models, when properly connected with human actions on land (land use and cover change-LUCC), can help identify areas that are most prone to desertification.

* e-mail: p.vlek@uni-bonn.de

Desertification

The issue of desertification has been hotly debated over the past few decades, in part because of the ambiguity regarding the definition of desertification in the current literature. Interpretations range from the advance of the desert boundary resulting from a drying climate to the development of desert-like conditions occurring in populated areas in semiarid but non-desert environments. Most contemporary definitions contain four important criteria as features of desertification:

(1) The condition pertains to arid, semiarid and dry sub-humid territories. or drier conditions
(2) The condition is caused chiefly by human action
(3) Land degradation is the underlying process
(4) The result is a persistent decline of productivity of useful biota to man and his animal support system.

According to these criteria, desertification is a condition that may occur in the absence of climate change. However, dryer and more irregular rainfall patterns, a characteristic shared by regions susceptible to desertification, would tend to aggravate the situation by increasing the pressure on land.

Especially, acute are the problems encountered by societies subsisting in arid and semi-arid regions, where droughts are frequent and the environment is highly vulnerable to degradation. In such regions, denudation of watersheds by deforestation, inappropriate cultivation methods and overgrazing inevitably destroys biodiversity. Denudation also unleashes processes of erosion by wind during dry periods and by water during brief torrential rainstorms. Consequently, the thin and vulnerable mantle of fertile topsoil is stripped off, thus exposing the infertile subsoil or even the bedrock. The loose material is eventually deposited as sediment along the beds, banks, and estuaries of rivers, where it tends to clog up the outflow and cause the formation of malaria-infested marshes.

Another process of land degradation typically comes into play where irrigation is introduced in the river valleys of semi-arid or arid regions. The traditional mode of irrigation in such areas is to divert water from the river (or from a reservoir created by damming the river) and to convey it via unlined channels to flat basins in which the water can be impounded periodically. About one third of the irrigation water is lost before it reaches the fields. Another one third is lost after application by flooding, which remains the most common method of irrigation. The existing inefficient modes of irrigation generally causes the water-table to rise, until–after some time (decades, perhaps)–it invades the root zone from below and causes the soil to become waterlogged. In these circumstances, many crops suffer from lack of aeration, and processes of anaerobiosis begin, including denitrification and methane generation.

A related but perhaps more insidious consequence of waterlogging is soil salination. Capillary rise of groundwater from a high water table toward the soil surface, where it evaporates, results in the precipitation of the salts generally contained in the groundwater. As it becomes saline, the upper zone of the soil restricts the growth of many crop plants. In extreme cases, the soil is rendered sterile. Salinity can also develop due to over-exploitation of groundwater that promotes surfacing of harmful salts.

Still another hazard of irrigation development is the spread of water-borne diseases. Such diseases range from dysentery, cholera, and malaria to onchocersiasis (river blindness) and schistosomiasis (bilharzia). They affect many millions of people in irrigated areas of Africa, Asia, and South America.

Curbing desertification

Modern technologies may help solve or alleviate many of these problems. Contour planting and terracing, grassed strips, minimum tillage, stubble mulch management, intercropping and agroforestry can greatly reduce erosion in rain-fed farming areas. In irrigated areas, the lining of canals with impervious materials or, preferably, the conveyance of water in closed watertight conduits (pipes) can help prevent water losses and the spread of some diseases (e.g. bilharzia). Effective soil drainage, as well as treatment, safe disposal or safe reuse of drainage effluent can also help in the control of diseases, as well as in the prevention of salinity development.

Greatly improved irrigation methods have been developed in recent decades to help prevent salination and waterlogging, and to achieve greater water-use efficiency. Instead of flooding the soil at infrequent intervals with unmeasured and excessive volumes of water (that cause waterlogging), these methods are based on the high-frequency, low-volume application of water in precise quantities. Delivery of water follows crop- water requirements, which depend on the weather as well as on the physiological stage of crop growth. Such irrigation methods, including drip, trickle, and micro-sprayer irrigation, also permit the efficient application of nutrients via the water supply and hence the utilization of coarse sandy and gravelly soils formerly considered unsuitable for irrigation. Moreover, these methods often result into higher productivity for a unit volume of water. Also, they are effective measures against desertification.

The envisioned intensification of agricultural production, both in favorable conditions of rain-fed farming and in irrigated areas, can help to alleviate the excessive pressure on marginal lands. Such lands should be revegetated and kept in reserve e.g., as auxiliary pastures to be utilized in drought seasons. The guiding principle is that development must mean improvement (or at least conservation) of the resource base in the long run, not its destructive exploitation for short-run gain.

Although desertification is a universal phenomenon, the solutions to it are not. Hence, none of the possibilities mentioned can or should be applied as "off-the-shelf" technologies to the problems faced by local communities in semi-arid regions of developing countries. Rather, the principles of efficient conservation of soil and water must be adapted to the specific environment and socio-economic conditions that prevail in each case. This should be done by enlisting the cooperation, wisdom, and experience of local farmers and stakeholders and by relying as much as possible, on local materials and workmanship. Only thus can the fight against desertification become self-regenerating, self-sustaining and self-reinforcing.

More attention and care must be devoted to stabilizing lands subject to degradation, to respecting the carrying capacity and sustainability of land, to the efficient use of soil and water resources, and to understanding the types of remedial actions appropriate to the restoration of degraded land. There is much to learn from a Bedouin teacher who explained that the totality of a living community is more than the mere sum of the parts, because it includes their numerous and complex interactions, both synergistic and antagonistic (Hillel, 1994). We should seek such synergism in the appropriate application of scientific principles and modern technology, giving full consideration to local conditions and to the experience and preferences of local societies, toward alleviating the problems caused by the environmental degradation known as "desertification".

Discussion

The ensuing discussion revolved around four principal questions:
(1) Do we have the means to properly identify desertification, particularly at its onset, and to quantify the problem?
(2) Do we need to be quantitative before action can and will be taken to reverse the problem, and what lessons can we learn from the global-climate change debate?
(3) What are the main actions that should be taken to forecast the potential occurrence of land degradation, to arrest the process before it advances, as well as to remedy or reclaim lands that have already been degraded?
(4) What are the appropriate policies that can promote the control of land degradation, and how can such policies be instituted, communicated, and enforced?

Early identification of desertification or the constituting processes of soil degradation and climate change was seen as highly desirable due to the long lead times required for finding proper solutions and for applying the solutions that are most suitable – environmentally, technically, economically, and socially – for specific locations. The need for early warning indicators was recognized and a catalogue of tools identified, ranging from remote sensing and monitoring of vegetative cover and soil surface moisture and erosion to the isotopic analysis of soil carbon. Vegetation monitoring-both from the point of biomass productivity and species dynamics-is particularly important, although the lack of a firm baseline by which to assess a continuous process makes interpretation problematic. After all, we are only very late observers of a destructive process that in too many places has been occurring for a very long time. The complexity of the desertification process and its location-specific nature make early detection a difficult task.

The participants in the discussion generally agreed that public attention to desertification would be enhanced if there were a way to measure the potential costs of not attending to the problem. The global change issue itself underwent a long process of avoidance, hesitation, and debate before it became widely acknowledged, and it gained momentum only after modeling attempts were made that quantified some of the problems. Unfortunately, we have not so far been very convincing in the task of quantifying the monetary losses associated with soil degradation, except for those areas that are closely monitored due to their intensive management (e.g. irrigation schemes). A major effort should be made to assess the cost of desertification. Some countries, such as Algeria, have achieved some progress in that direction. To be effective, the task must be a multidisciplinary effort from the outset, involving climatology, ecology, hydrology, pedology, agronomy, sociology and possibly other areas of expertise as well. Mesoscale modeling of watersheds can help in forecasting where problems are likely to occur and *to* require early remediation.

Other action that could be taken immediately include applying improved technologies in known problem areas, seeking means of alleviating the pressures on the land, creating off-farm employment, shifting from emphasis on food self sufficiency to food security, and communicating effectively with policy makers as well as with stake holders. Some skepticism was expressed regarding the applicability of modern technologies to certain areas that are affected by desertification (e.g. Botswana). The answer may be that such technologies should not simply be transferred in fixed form from developed to developing countries. Rather these should be adapted or improved to fit the latter's specific circumstances (e.g. the relative availability of labor and local materials versus the scarcity of capital for purchasing equipment from abroad).

The group adjourned on a note of optimism that a scientifically integrated and politically coordinated attack on the desertification issue can indeed avert the brunt of the problem. Such an approach should definitely not be delayed until all monitoring systems have been put in place and the extent and costs of the problem have been fully and precisely defined. Rather, a proactive program should be started immediately.

References

Falkenmark M (1997) Meeting water requirements of an expanding world population. Phil Trans R Soc Lond 352: 929–936

Hillel D (1992) Out of the Earth: Civilization and the Life of the Soil. The University of California Press, Los Angeles

Hillel D (1994) The Rivers of Eden: The Struggle for Water and the Quest for Peace in the Middle East. Oxford University Press, New York

Subject Index

Printing and Binding: Stürtz AG, Würzburg